Concise Guide to Heat Exchanger Network Design

Xian Wen Ng

Concise Guide to Heat Exchanger Network Design

A Problem-based Test Prep for Students

 Springer

Xian Wen Ng
Singapore, Singapore

ISBN 978-3-030-53500-1   ISBN 978-3-030-53498-1   (eBook)
https://doi.org/10.1007/978-3-030-53498-1

This Springer imprint is published by the registered company Springer Nature Switzerland AG
The registered company address is: Gewerbestrasse 11, 6330 Cham, Switzerland

Preface

Thermal energy is one of the most important sources of energy driving processes in industrial plants, and one of the key engineering outcomes is often effective heat integration and energy recovery between process streams. It is critical for engineers and related professionals to master the technique of designing optimal heat exchanger networks that achieve maximum energy conservation, while not compromising on desired production objectives.

This pocket guide serves as a supplementary learning resource that will help students deconstruct some of the most challenging problems encountered in the topic of Heat Exchanger Networks. The practice problems in this book were carefully selected to represent the most commonly encountered problems found in tests and examinations. With the comprehensive worked solutions and detailed explanations provided for each problem, students will be able to closely follow the thought process of problem-solving from start to finish, thereby hone their skills in applying abstract theoretical concepts to solving practical problems which is critical for acing examinations.

The mix of both numerical and open-ended problems included in this book will help students gain a well-rounded understanding of Heat Exchanger Networks and related design principles. With the use of this study guide, students will become proficient not only in handling numerical analysis but also in relating the significance of desktop problem-solving to the larger real-world context.

Singapore, Singapore   Xian Wen Ng

Acknowledgements

My heartfelt gratitude goes to the team at Springer for their unrelenting support and professionalism throughout the publication process. Special thanks to Michael Luby, Brian Halm, and Hema for their constant effort and attention towards making this publication possible. I am also deeply appreciative of the reviewers of my manuscript who had provided excellent feedback and numerous enlightening suggestions to help improve the book's contents.

Finally, I wish to thank my loved ones who have, as always, offered only patience and understanding throughout the process of making this book a reality.

Contents

1 Fundamentals of Heat Integration   1

2 Energy Cascade and Pinch Analysis 19

3 Euler's Theorem and Grand Composite Curves 63

4 Complex Hen Design Problems 109

Index 147

About the Author

Xian Wen Ng graduated with First-Class Honors from the University of Cambridge, UK, with a Master's Degree in Chemical Engineering and Bachelor of Arts in 2011 and was subsequently conferred a Master of Arts in 2014. She was ranked second in her graduating class and was the recipient of a series of college scholarships including the Samuel Taylor Marshall Memorial Scholarship, Thomas Ireland Scholarship, and British Petroleum Prize in Chemical Engineering for top performance in consecutive years of academic examinations. She was also one of the two students from Cambridge University selected for the Cambridge-Massachusetts Institute of Technology (MIT) exchange program in Chemical Engineering, which she completed with Honors with a cumulative GPA of 4.8 (5.0). During her time at MIT, she was also a part-time tutor for junior classes in engineering and pursued other disciplines including economics, real estate development, and finance at MIT and the John F. Kennedy School of Government, Harvard University. Upon graduation, she was elected by her college fellowship to the title of scholar, as a mark of her academic distinction.

Since graduation, she has been keenly involved in teaching across various academic levels. Her topics of specialization range from secondary-level mathematics, physics, and chemistry up to tertiary-level mathematics and engineering subjects. Some of her recent publications include *Engineering Problems for Undergraduate Students* and *Pocket Guide to Rheology*, which are practice books similarly written for students taking engineering and related STEM courses at higher education and university levels. These books aim to sharpen students' problem-solving skills and put them in good stead for tests and examinations.

Chapter 1
Fundamentals of Heat Integration

Problem 1

Consider an industrial process that comprises three process streams as tabulated below. Determine if the overall system is a net heat source or heat sink.

Process stream	Supply temperature, T_{supply} [°C]	Target temperature, T_{target} [°C]	Flow rate heat capacity, W [kW/K]
1	200	70	0.4
2	100	170	0.7
3	80	150	1.0

Solution 1

Classifying Process Streams—Hot and Cold Streams

Before we begin, we should first identify the nature of the process streams in terms of whether they are "hot" or "cold" streams. This is a key step in tackling heat exchanger network problems, especially so when problems get more complex.

From the table given, we note that one of the streams is hot while the other two are cold streams. The hot stream needs to be cooled down to a prescribed target temperature, while the cold streams need to be heated to their target temperatures, in order to meet required processing conditions.

Process Stream	Hot or cold?	Supply Temperature, T_{supply} [°C]	Target Temperature, T_{target} [°C]	Flow rate heat capacity, W [kW/K]
1	Hot	200	70	0.4
2	Cold	100	170	0.7
3	Cold	80	150	1.0

© The Editor(s) (if applicable) and The Author(s), under exclusive license to Springer Nature Switzerland AG 2021
X. W. Ng, *Concise Guide to Heat Exchanger Network Design*,
https://doi.org/10.1007/978-3-030-53498-1_1

Note also that "flow rate heat capacity" denoted as W is simply heat capacity applied to continuous flows such as process streams in steady-state operations. Mathematically, it is the product of heat capacity and flow rate as shown below, whereby Watt is equivalent to Joules per second.

$$\text{Flow rate heat capacity } [\text{kW/K}] = \text{Flow rate } [\text{kg/s}] \times \text{Heat capacity } \left[\frac{\text{kJ}}{\text{kg} \cdot \text{K}}\right]$$

Total Heating (or Cooling) Requirement

In this simple example, we can calculate the heating (or cooling) requirement for the entire process (i.e., heating and cooling duties required) in order to achieve the desired target temperatures of the process streams.

The total required cooling duty for the hot stream 1 can be determined as follows:

$$Q_c = \left(T_{\text{supply}} - T_{\text{target}}\right) \times W_1 = (200 - 70) \times 0.4 = 52 \text{ kW}$$

The total required heating duties for cold streams 2 and 3 can be determined as follows:

$$Q_{\text{H},2} = \left(T_{\text{target}} - T_{\text{supply}}\right) \times W_2 = (170 - 100) \times 0.7 = 49 \text{ kW}$$

$$Q_{\text{H},3} = \left(T_{\text{target}} - T_{\text{supply}}\right) \times W_3 = (150 - 80) \times 1.0 = 70 \text{ kW}$$

$$Q_{\text{H}} = Q_{\text{H},2} + Q_{\text{H},3} = 49 + 70 = 119 \text{ kW}$$

The net heating requirement is therefore the sum of all heating and cooling needs as computed below. Assuming we define $Q > 0$ as a system that requires input of heat, then we add a negative sign to the cooling duty and arrive at the following.

$$Q = -Q_c + Q_{\text{H}} = -52 + 119 = 67 \text{ kW}$$

This shows that the overall system requires a heating duty of 67 kW, and it is therefore also a ***net heat sink*** (system needs to "take in" heat and the supply of this heat is done through the heat exchanger which provides the heating duty of 67 kW).

Problem 2

Discuss the concept of heat integration in relation to the heating (or cooling) requirements in a typical process plant. Consider any benefits and trade-offs of heat integration and explain how heat exchangers may help achieve heat integration.

Solution 2

Why the Need for Heat Integration?

Heat energy is an indispensable source of energy driving many industrial processes, thereby ensuring the optimal use of heat through energy conservation and recovery techniques. The effective use of heat is therefore a critical consideration for most process or chemical engineers as it ensures cost-effective process operations.

Heat exchangers (HXers) are essential unit operations in most processing plants as they supply or remove heat where necessary, to maintain stable operating conditions. Hence, it makes sense that the optimal design of heat exchanger networks (HENs) is pertinent to achieving effective heat use via integration and optimal energy conservation.

Considerations in HEN Design

The main outcomes in designing HENs may be summarized as follows:

1. Fulfill process requirements

   • Process streams need to reach target temperatures via heating or cooling from initial (or supply) temperatures.

2. Minimize costs

   • *Utility costs* of heating and/or cooling duties.

     – The cost of adding heating or cooling utilities can be minimized through heat integration between process streams.
     – By using pre-existing hot process streams (which require cooling to reach target temperatures) to heat up cold process streams (which require heating to reach target temperatures), we minimize the need for external heating/cooling utilities. Effectively, we are making use of a "purposeful interaction" between process streams themselves (through HEN design) to fulfill the streams' individual heating or cooling needs.

   • *Capital costs* of HXers

     – Cost can be minimized if smaller or fewer HXers are used. Therefore, capital cost is a trade-off in heat integration since heat integration inherently requires the use of HXers. This trade-off can be balanced with savings in utility costs achieved with better heat integration and hence energy recovery.

Illustrated Example of Heat Integration

Consider a simple processing facility with two streams as shown below, one hot and one cold.

Process stream		Supply temperature, T_{supply} [°C]	Target temperature, T_{target} [°C]	Flow rate heat capacity, W [kW/K]
1	Hot	250	58	0.31
2	Cold	110	170	1.2

We can conduct simple heat integration by using stream 1 to heat up stream 2 using a HXer that places both streams in close proximity. This is illustrated in the diagram below, where we denote the duty of this HXer as Q, and the loads of any additional heating and cooling utilities required as Q_H and Q_C respectively.

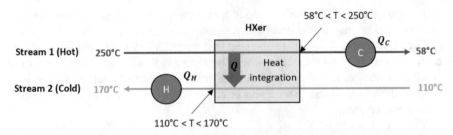

As both streams pass through the HXer in a countercurrent manner, stream 1 gets cooled down while stream 2 gets heated up. Both streams end up with temperatures closer to their required target temperatures upon exiting the HXer. If the streams are able to reach their respective target temperatures via this process, then no additional utilities would be required. However, where there are unmet heating or cooling loads, i.e., if target temperatures are not reached, then additional heating or cooling utilities can be added to completely fulfill required heating and cooling loads of the system. These additional or external utilities are shown as shaded circular symbols in the diagram above.

We can further analyze our system in a simple plot where temperature is plotted against enthalpy. We can think of enthalpy simply as heat energy. There are a few reasons why this graph is useful.

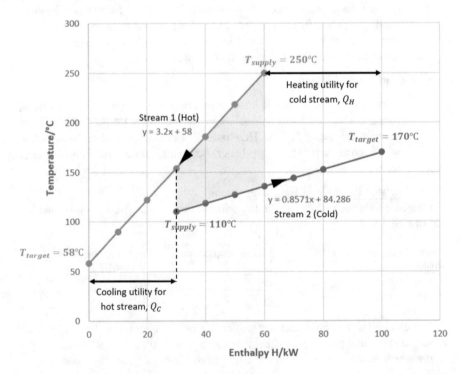

Let us explore some of its key features.

- **Hot and cold streams**

  - The hot stream is represented by the orange line, and it lies above the line marked in blue representing the cold stream.
  - The direction along which stream temperatures change goes from supply temperature to target temperature, as shown in black arrows.

- **Heat integration (or heat transfer) region**

  - The region shaded in orange corresponds to where both lines "overlap" vertically. This is also the heat integration or heat transfer region, where we may use a hotter stream (stream 1) to heat up a colder stream (stream 2) using a HXer that brings both streams in close proximity. Through this heat transfer, both streams move closer to their target temperatures.

- **External utilities**

  - As the temperature of the hot stream exiting the HXer has not yet reached its target temperature (though it is now closer to the target temperature than if without the HXer), an external cooling utility needs to be added in order to reach the target temperature and completely fulfill the hot stream's cooling requirement. Graphically, the additional cooling load fulfilled using external utilities is represented by the region marked Q_C.
  - In the same way, the additional heating utility required for the cold stream to fulfill its remaining heating load is represented as Q_H.

- **Flow rate heat capacity**

  - Note that the gradient (or slope) of the lines is simply the reciprocal of heat capacity of the streams. Since the streams are in continuous flow, their heat capacities are referred to as "flow rate heat capacities."

    Flow rate heat capacity = Mass flow rate × Specific heat capacity

$$W \left[\frac{kW}{K} \right] = W \left[\frac{kJ}{s \cdot K} \right] = \dot{m} \left[\frac{kg}{s} \right] \times c_p \left[\frac{kJ}{kg.K} \right]$$

$$\text{Gradient} = \frac{\text{Temperature [K]}}{\text{Enthalpy [kW]}} = \frac{1}{W}$$

  - If the lines were curves, then flow rate heat capacity is not constant and can be determined by finding the tangent to a specific point of interest on the curve.
  - We can derive straight-line equations for each of the lines, starting from the following expression that relates enthalpy (or heat) Q to temperature change ΔT.

$$Q = W \cdot \Delta T$$

For process streams, we will usually know the temperatures entering the HXer, but need to compute outlet temperatures. This can be done by applying the above equation. For the hot stream,

$$T_{in} - T_{out} = \frac{1}{W_H}(Q)$$

$$T_{out} = T_{in} - \frac{Q}{W_H}$$

The straight-line equation for the hot stream above uses T_{in} and T_{out} as defined points (marked as red dots below) on the hot stream line in orange. Similarly, the straight-line equation for the cold stream uses T_{in} and T_{out} as defined points on the cold stream line (in blue).

$$T_{out} - T_{in} = \frac{1}{W_C}(Q)$$

$$T_{out} = T_{in} + \frac{Q}{W_C}$$

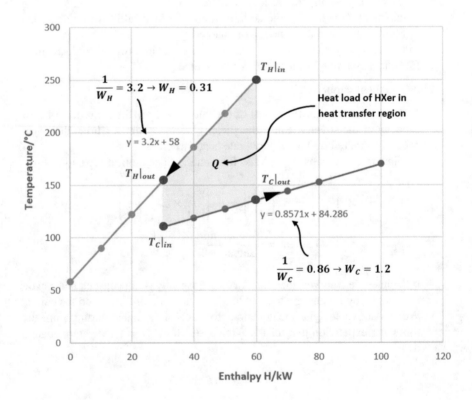

Problem 3

Discuss the design considerations for heat exchanger units using suitable diagrams where appropriate. You may consider the key words suggested below.

- **Design equation for heat exchanger**
- **Degree of freedom**
- **Maximum energy recovery/minimum energy requirement**
- **Pinch point**
- **Minimum approach temperature**
- **Second law of thermodynamics**

Solution 3

Design Equation for Heat Exchanger (HXer)

Consider a simple HXer unit with two streams, one hot and one cold, as illustrated below. The objective here is to design a HXer that achieves optimal heat integration and energy recovery between the streams, thereby allowing both streams to achieve their target temperatures at minimal utility cost.

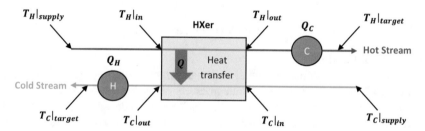

The descriptions of the notations used in the above diagram are shown below.

Notations	Descriptions
$T_H|_{target}$, $T_C|_{target}$	Target temperatures for hot and cold streams (or final desired temperatures)
$T_H|_{supply}$, $T_C|_{supply}$	Supply temperatures for hot and cold streams (or initial starting temperatures)
$T_H|_{in}$, $T_C|_{in}$	Inlet temperatures of hot and cold streams entering HXer
$T_H|_{out}$, $T_C|_{out}$	Outlet temperatures of hot and cold streams exiting HXer

The design equation for HXers originates from the fundamental principle of heat transfer and is shown below. We notice the presence of a heat transfer coefficient, a heat transfer area, and a temperature difference (driving force for heat transfer) in this equation. Q [J/h] is the rate of heat transfer between two process streams flowing through the HXer.

$$Q = UA\Delta T_{LM}$$

In this equation, U $\left[\frac{J}{h \cdot m^2 \cdot K}\right]$ denotes the overall heat transfer coefficient while A [m^2] represents the heat transfer area. ΔT_{LM} [K] represents an average (also known as the log mean average) temperature difference between the inlet and outlet temperatures of both streams and can be computed as shown below.

$$\Delta T_{LM} = \frac{\left(T_H|_{in} - T_C|_{out}\right) - \left(T_H|_{out} - T_C|_{in}\right)}{\ln\left(\frac{T_H|_{in} - T_C|_{out}}{T_H|_{out} - T_C|_{in}}\right)}$$

The log mean average is suitable for HXer units since it caters for the fact that there are two streams involved, whereby the temperatures of both streams vary as they pass through the HXer unit.

Points to note:

Heat transfer has to occur from a hotter region to a colder region; therefore the following has to be true:

$$T_H|_{in} > T_C|_{out}, \quad T_H|_{out} > T_C|_{in}$$

We know as well that the hot stream loses heat while the cold stream gains heat as they pass through the HXer. Therefore, the following is also true:

$$T_H|_{in} > T_H|_{out}, \quad T_C|_{out} > T_C|_{in}$$

Degree of Freedom

The degree of freedom is essentially a design parameter that we are able to change in order to achieve different design outcomes. By observing the design equation, we notice that the parameter that is most "flexible" for us to design (i.e., adjust to our needs) is heat transfer area A since it can be changed by varying the size and/or numbers of HXers used.

It is therefore useful to understand the correlation between the extent of heat integration between hot and cold streams and heat transfer area (hence HXer size). The extent of heat integration also directly affects the addition and/or sizing of external utilities where required.

Let us first consider a base case as shown below, with the following data provided.

Base case:

Process stream	T_{supply} [°C]	T_{target} [°C]	Flow rate heat capacity, W [kW/K]	Utilities required, Q_H or Q_C [kW]
Hot	250	58	0.3125	Cooling of 30 kW
Cold	110	170	1.167	Heating of 40 kW

We can plot the operating lines of both streams on a temperature against enthalpy (or heat) graph. The **end points of the lines are defined based on the supply and target temperatures** provided above. In addition, the **gradients of the lines are defined** since they are equivalent to the reciprocals of the flow rate heat capacities as shown below.

$$\text{Gradient of hot stream line} = \frac{1}{W_H} = \frac{1}{0.3125} = 3.2 \text{ K/kW}$$

$$\text{Gradient of cold stream line} = \frac{1}{W_C} = \frac{1}{1.167} = 0.857 \text{ K/kW}$$

Inter-Stream Heat Transfer (Heat Integration)

After plotting both the hot and cold stream lines, we observe an area of overlap between the two lines shown by the shaded region. This represents the heat exchange or heat integration region where a HXer unit that brings the hot and cold streams in close proximity allowed for inter-process stream heat transfer. Let us denote the amount of heat transferred in the HXer for the base case as Q_{base} as shown in the plot below.

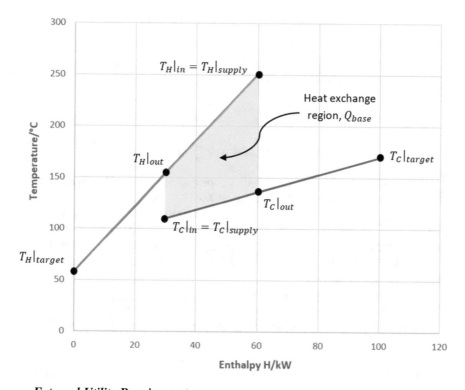

External Utility Requirements

We can also find external utility requirements from the graph. The **external heating utility** required to raise the temperature of the cold stream further (upon its exit from the HXer where it gained some heat from the hot stream) to its final target temperature is $Q_{H,base} = 40$ **kW**. The **external cooling utility** required to further lower the temperature of the hot stream (upon its exit from the HXer where it lost some heat to the cold stream) down to its final target temperature is $Q_{C,base} = 30$ **kW**.

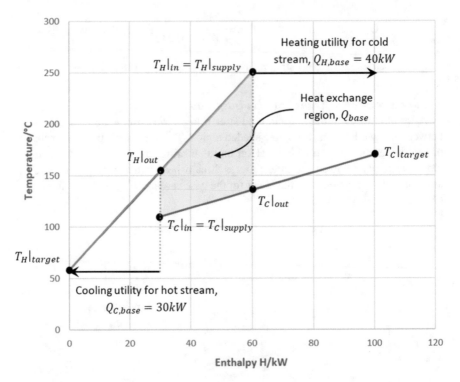

Let us now explore various case studies depicting scenarios with varying extents of heat integration as compared to the Base Case.

Assuming that we keep the supply and target temperatures as well as flow rate heat capacities of both streams the same, we will now explore the effect of **shifting the cold stream line.** This effectively changes the **extent of heat integration** between the two streams, and hence our HXer design. The key points of the various cases are examined below.

Case 1: Theoretical Maximum Heat Exchange ($\Delta T_{min} = 0$)

Starting from the base case, if we now shift the cold stream line further to the left, it will eventually reach a point when it **just touches the hot stream line**. This is shown below where the original heat exchange area for the base case Q_{base} is marked in dotted lines, while the **orange shaded area Q_1 refers to the new heat exchange region (larger area)** for Case 1. We note that with this shift, the **heating and cooling utility duties are also both reduced** from the base case and are in fact at their lowest possible values. By reading off our plots on a scaled graph, we can obtain $Q_{C, \text{ case 1}}$ to be approximately 16.25 kW and $Q_{H, \text{ case 1}}$ to be approximately 26.25 kW.

$$Q_{C,\text{case 1}} = 16.25 \text{ kW} < Q_{C,\text{base}} = 30 \text{ kW}$$

$$Q_{H,\text{case 1}} = 26.25 \text{ kW} < Q_{H,\text{base}} = 40 \text{ kW}$$

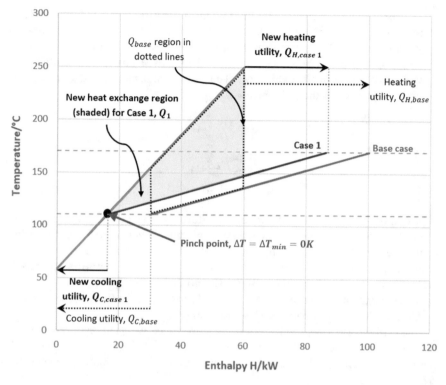

There are several key points to note for Case 1:

- This case represents the ideal (thermodynamic limit) scenario of **maximum heat exchange** or **maximum energy recovery**. Both streams are most "heat integrated" in this configuration because the **region of overlap between the two lines is maximized**.

- This state corresponds to requiring **minimum utility duties**, and this is often abbreviated as **MER** (maximum energy recovery or minimum energy requirement). It is common for design engineers to work towards achieving MER for HEN design.

- The Second Law of Thermodynamics states that heat energy must flow from a hotter region (higher temperature) to a colder region (lower temperature); this means that the **heat transfer direction on the graph must point vertically downwards** from the **hot stream line to the cold stream line**.

  - This is also why the cold stream line cannot move any further left from touching the hot stream line.
  - If the cold stream line were to cross the hot stream line, part of the cold stream line would be above the hot stream line and this would mean heat travels from cold to hot in the vertical downwards direction, which is not possible.

Case 1 depicts a theoretical limit or ideal scenario that is not achievable in real life. This is because in order to have a pinch point where the hot and cold stream lines **meet**, the temperature difference at that point must be zero (i.e., $\Delta T_{min} = 0$) which means that the heat exchange **area has to be infinitely big, which is not practically achievable**. This can be mathematically shown from the design equation for the HXer below.

$$Q = UA\Delta T_{LM}$$

$$A = \frac{Q}{U\Delta T_{LM}}$$

$$\Delta T_{min} \rightarrow 0, \ A \rightarrow \infty$$

We know that in reality, **HXers are of finite size**. Therefore, when designing actual HXers, we will need to stipulate a minimum temperature difference that cannot be zero (as in Case 1, the theoretical limit) but a small positive non-zero value ($\Delta T_{min} > 0$). This value of ΔT_{min} is also known as the **minimum approach temperature** and is commonly **between 10 and 20 K**.

We will explore this scenario ($\Delta T_{min} > 0$) in our next case study, Case 2 below.

Case 2: Practical Maximum Heat Exchange ($\Delta T_{min} \geq 0$)

The earlier Case 1 depicted an ideal scenario, which we mentioned was impractical in real life. In reality, the temperature difference at the pinch should be a small positive non-zero value. Let us assume an arbitrary minimum value for $\Delta T_{min} = 15$ K.

Recalling that Case 1 represented the left-most position that the cold stream line could take, we may deduce that the now-positive value of ΔT_{min} will mean that the cold stream line will be horizontally shifted to a position in between the base case and Case 1, such that the smallest vertical gap between the hot and cold stream lines is now 15 K.

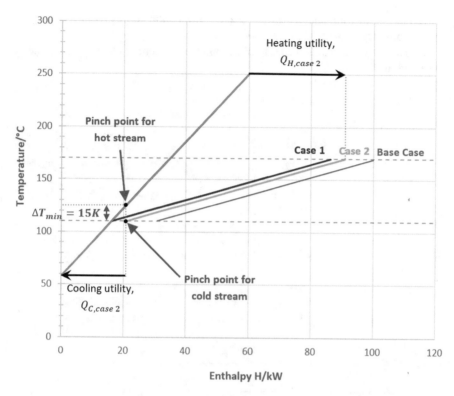

From a scaled plot, we may read off the utility values as shown below.

$$Q_{C,\text{case } 2} = 20.94 \text{ kW}$$

$$Q_{H,\text{case } 2} = 30.94 \text{ kW}$$

We note that $Q_{C,\text{ case } 1} < Q_{C,\text{ case } 2}$ and $Q_{H,\text{ case } 1} < Q_{H,\text{ case } 2}$ make sense since Case 1 represents the ideal scenario where the least amount of external utility is required (theoretical limit). As we increase $\Delta T_{\min}$ from zero to a non-zero value in Case 2 (although still a small positive value), the external utility duties will also increase as a result.

We also note that $Q_{C,\text{ case } 2} < Q_{C,\text{ base case}}$ and $Q_{H,\text{ case } 2} < Q_{H,\text{ base case}}$ implies that the system in the base case has further scope to increase the level of heat integration between process streams in order to reach the state in Case 2, where we have maximum/optimum inter-stream heat exchange and correspondingly the minimum external utility duties required.

Demonstrating the First Law of Thermodynamics

For Case 2 (maximum heat exchange that is practically achievable), consider how the plots shifted from their positions in Case 1 to Case 2 and implications on the cooling utility Q_C required. We notice that as we go from Case 1 to Case 2, we moved the cold stream line to the right, and the new cooling utility required, $Q_{C,\,case\ 2}$ increased by an amount. Let us denote this increase as ΔQ_c.

We can relate the value of ΔQ_c to ΔT_{min} and the gradient of the hot stream line as observed in the graph below.

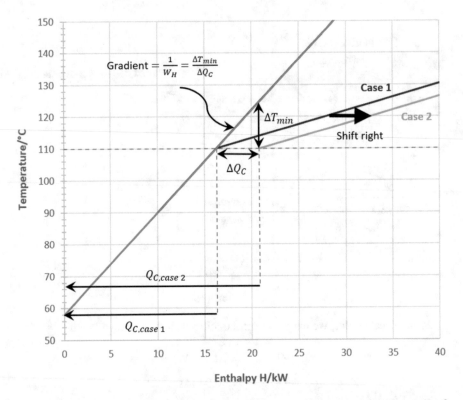

Recalling that the gradient of the stream lines is equivalent to the reciprocals of their heat capacities W, we can then compute the value of ΔQ_c and subsequently determine $Q_{C,\,case\ 2}$.

$$\Delta Q_c = W_H \Delta T_{min} = 0.3125(15) = 4.69 \text{ kW}$$

$$Q_{C,case\ 2} = Q_{C,case\ 1} + \Delta Q_c = 16.25 + 4.69 = 20.94 \text{ kW}$$

Now that we have found the cooling utility duty $Q_{C,\,case\ 2}$, let us look at the top right corner of the graph to see what happens to the heating utility as we go from Case 1 to 2.

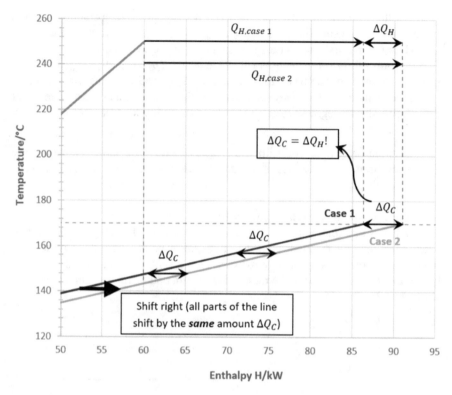

Figuring out the value of $\Delta Q_{\text{H, case 2}}$ is relatively straightforward after we have figured out ΔQ_c from earlier. We may observe easily from the diagram above that all parts of the cold stream line shift horizontally by the **same amount** of $\Delta Q_c = 4.69\,\text{kW}$. Therefore, when we trace the cold stream line to the other end at the top right, we notice that the heating utility should also increase by the same amount of 4.69 kW.

$$Q_{\text{H,case 2}} = Q_{\text{H,case 1}} + \Delta Q_c = 26.25 + 4.69 = 30.94\ \text{kW}$$

This is in fact a key observation, as it demonstrates the First Law of Thermodynamics, which states that energy is conserved. With no changes made to the system (process streams have the same supply and target temperatures even as we vary the degree of heat integration), the overall energy balance of the system should be maintained. This means that if we **increase** the heating utility (an energy input to the system), the corresponding increase in cooling utility will be **equivalent** to ensure energy conservation.

To conclude, let us summarize key points for Case 2:

- In reality where $\Delta T_{\min} \neq 0$, pinch temperatures come as a pair, one for the hot stream and one for the cold stream. This is different from the scenario in Case 1 where the pinch temperature was equivalent for both streams.

- When the temperature difference between the two pinch temperatures is at the minimum achievable (i.e., $\Delta T|_{\text{pinch}} = \Delta T_{\text{min}} = 15$ K), we have **maximum heat exchange** or **maximum energy recovery**. This also corresponds to having **minimum utility requirements** which we denote $Q_{C,\text{ min}}$ and $Q_{H,\text{ min}}$.
- Note that in the earlier base case, $\Delta T|_{\text{pinch}} > \Delta T_{\text{min}}$. This explains why the utility requirements in the base case were not at their minimum, and the system could be further heat integrated to achieve even lower utility loads as demonstrated in Case 2.

  - In summary, $Q_{C,\text{ case 1}}$ (ideal) $< Q_{C,\text{ case 2}}$ (MER) $< Q_{C,\text{ base case}}$ (general case with further scope for heat integration). This trend is likewise for Q_H values.
  - Also, $Q_{C,\text{ case 2}}$ (MER) $= Q_{C,\text{ min}}$. Likewise, for Q_H values.

Case 3: No Heat Exchange ($Q = 0$)

Finally, let us explore another limiting case where we shift the cold stream line much more towards the right until it just reaches a point where there is **no heat integration**. From this point onwards (i.e., as we shift the cold stream line even further to the right), there will be **no overlap region** and therefore inter-stream heat exchange, $Q = 0$.

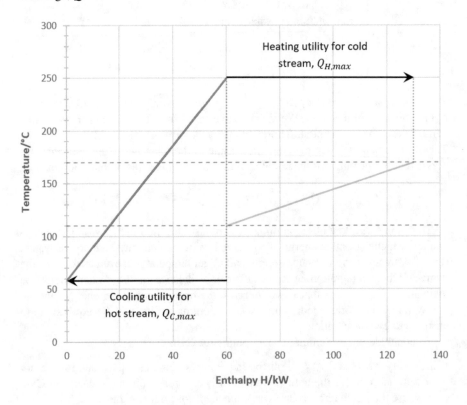

As before, there are some key takeaways for Case 3:

- There is no heat exchange, hence no pinch point nor minimum approach temperature (irrelevant).
- This case represents **no heat exchange** or **no energy recovery** and therefore corresponds to **maximum utility requirements**.

Summary of all Cases (Base Case, Cases 1–3)

- To summarize all cases, we have $Q_{C, \text{ case 1}} < Q_{C, \text{ case 2}} < Q_{C, \text{ base case}} < Q_{C, \text{ case 3}}$ and this trend is likewise for Q_H values.
- $Q_{C, \text{ case 2}} = Q_{C, \text{ min}}$ and likewise for Q_H values.
- $Q_{C, \text{ case 3}} = Q_{C, \text{ max}}$ and likewise for Q_H values.

A summary table is shown in the next page.

HEN design features	Ideal case (theoretical limit)	Practically achievable cases		
	Case 1	Case 2	Base case	Case 3
Pinch temperatures	• Single temperature • $T_{\text{pinch,hot}} = T_{\text{pinch,cold}}$	• Two temperatures ($T_{\text{pinch,hot}}$, $T_{\text{pinch,cold}}$) • $T_{\text{pinch,hot}} > T_{\text{pinch,cold}}$		N.A.
Minimum approach at pinch	Zero	• $\Delta T = \Delta T_{\text{min}} = T_{\text{pinch,hot}} - T_{\text{pinch,cold}}$ • Minimum possible ΔT, ~10–20 K	• $\Delta T = T_{\text{pinch,hot}} - T_{\text{pinch,cold}}$ • Not the minimum possible, $\Delta T > \Delta T_{\text{min}}$	
Utility duties, Q_c and Q_H	Minimum (theoretical)	• Minimum (achievable), denoted $Q_{C,\text{min}}$, $Q_{H,\text{min}}$ • Larger than ideal Case 1, smallest out of all practically achievable cases	• Larger than $Q_{C,\text{min}}$, $Q_{H,\text{min}}$ • Can further minimize through greater heat integration	Maximum, largest out of all cases
Energy recovery HXer duty, Q	Maximum (theoretical)	Maximum (achievable)	Moderate, can be increased further	None
Heat integration				

Chapter 2
Energy Cascade and Pinch Analysis

Problem 4

Consider a simple industrial process where two feed streams are first converted into an intermediate product via a reaction, then distilled to obtain final product streams in the distillate and bottoms. The unit operations in this process are illustrated in the diagram below. The numbers in purple indicate stream temperatures, while the numbers in brackets indicate heat capacities (in kW/K) of the respective streams.

In the distillation column, superheated steam at 220 °C provides heat to the reboiler, while the condenser takes in cooling water at 25 °C for its cooling duty and cooling water leaves the condenser with an exit temperature of 35 °C.

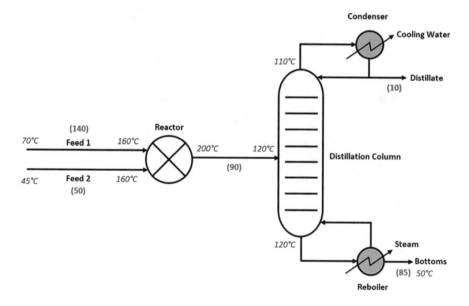

X. W. Ng, *Concise Guide to Heat Exchanger Network Design*,
https://doi.org/10.1007/978-3-030-53498-1_2

Given that the reboiler and condenser duties are both 3.2 MW,

(a) **Tabulate stream data for the process, indicating all the information required to design a heat exchanger network for heat integration between the streams.**
(b) **Using a minimum approach temperature $\Delta T_{min} = 20$ K, determine the pinch point and minimum utility duties using an energy cascade diagram.**
(c) **Explain the significance of the pinch point in relation to the terms "heat source" and "heat sink." Comment on the implications of "heat leakage" on external utilities required.**

Solution 4

(a)

Unlike Problem 3, this problem now considers multiple cold and hot streams. Hence, we will expand the earlier concept of individual hot and cold streams to hot and cold **composites** (sum of multiple streams).

Labeling of Streams (Hot or Cold)

When we have multiple streams, it is good practice to first **label the streams and identify them as either hot or cold streams**. This is because we will later need to combine individual hot (or cold) streams within the same temperature intervals to form a "hot (or cold) composite".

In the diagram below, we have added labels 1–6 for the six individual streams in our system.

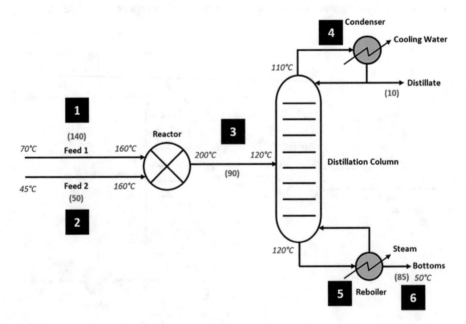

Process Streams

To identify the streams as hot or cold, we check their supply and target temperatures, T_{supply} and T_{target}, taking note that the streams begin with their supply temperatures and end with their target temperatures. Therefore,

If $T_{supply} > T_{target}$, it is a hot stream

If $T_{supply} < T_{target}$, it is a cold stream

From the known supply and target temperatures of each process stream, we know that hot streams are streams 3 and 6, while cold streams are streams 1 and 2.

Reboiler and Condenser

We know that the condensation process occurring in the condenser is exothermic, hence the stream "loses heat" like a hot stream as it changes phase from vapor to liquid at a constant temperature (supply and target temperatures are equal at 110 °C). Similarly, for the reboiler, the boiling process is endothermic, hence the stream "absorbs heat" like a cold stream as it changes phase from liquid to vapor at the constant temperature of 120 °C. Hence, we mark the reboiler stream 4 as hot and condenser stream 5 as cold.

Stream label	Stream name	T_{supply} [°C]	T_{target} [°C]
1 (cold) or "1C"	Feed 1	70	160
2 (cold) or "2C"	Feed 2	45	160
3 (hot) or "3H"	Reactor Product	200	120
4 (hot) or "4H"	Condenser	110	110
5 (cold) or "5C"	Reboiler	120	120
6 (hot) or "6H"	Bottoms Product	120	50

Enthalpy Loads of Streams

Process Streams

The flow rate heat capacities (W) for the streams were also given in the problem and can be included in the table for each stream. We will also calculate the respective enthalpy loads (ΔH) of the process streams, given by:

$\Delta H = W(T_{supply} - T_{target})$ for hot streams

$\Delta H = W(T_{target} - T_{supply})$ for cold streams

Enthalpy load also means the amount of heating or cooling needed for each stream to reach their target temperatures. Therefore for our hot streams, we have:

$$\Delta H_3 = W_3(200 - 120) = 90(80) = 7200 \text{ kW}$$

$$\Delta H_6 = W_6(120 - 50) = 85(70) = 5950 \text{ kW}$$

And for our cold streams, we have:

$$\Delta H_1 = W_1(160 - 70) = 140(90) = 12600 \text{ kW}$$

$$\Delta H_2 = W_2(160 - 45) = 50(115) = 5750 \text{ kW}$$

Reboiler and Condenser

We know that the condenser and reboiler duties are 3.2 MW each. This value also represents their enthalpy loads, which can be added directly into the table.

We may now update and complete our stream table with the heat capacities and enthalpies found.

Stream label	Stream name	T_{supply} [°C]	T_{target} [°C]	Flow rate heat capacity, W [kW/K]	ΔH [kW]
1 (cold) or "1C"	Feed 1	70	160	140	12600
2 (cold) or "2C"	Feed 2	45	160	50	5750
3 (hot) or "3H"	Reactor Product	200	120	90	7200
4 (hot) or "4H"	Condenser	110	110	–	3200
5 (cold) or "5C"	Reboiler	120	120	–	3200
6 (hot) or "6H"	Bottoms Product	120	50	85	5950

(b)

It is common to encounter HEN design problems requiring students to **find pinch temperatures** for optimal heat integration. One of the most useful methods is using the **energy cascade diagram**.

Let us summarize the steps involved in constructing an energy cascade. By remembering these simple steps, you will be able to tackle most heat integration problems in tests and exams.

Steps for Drawing an Energy Cascade Diagram

(a) **Write down all the supply and target temperatures for the hot and cold streams,** then arrange them in **descending order**. In this problem, we have the following hot stream temperatures in descending order. [Note that 110 °C is for the condenser and is marked with an asterisk (*) for special treatment later.]

Hot:	200°C	120°C	*110°C	50°C

(b) From the hot stream temperatures from part (a) above, we now **create corresponding cold stream temperatures** based on $\Delta T_{min} = 20$ K and knowing that $\Delta T_{min} = T_{hot} - T_{cold}$. The results are shown below.

Hot:	200°C	120°C	*110°C	50°C
Cold:	180°C	100°C	*90°C	30°C

(c) We now repeat steps (a) and (b) for the cold streams. [*Similar to the condenser in part (a), 120 °C is for the reboiler and is marked in asterisk (*) to be specially treated later.]

The cold process streams have these temperatures in descending order:

| Cold: | 160°C | *120°C | 70°C | 45°C |

The corresponding hot stream temperatures are similarly found using $\Delta T_{min} = 20$ K.

| Cold: | 160°C | *120°C | 70°C | 45°C |
| Hot: | 180°C | *140°C | 90°C | 65°C |

(d) Combining all temperatures found from parts (a) to (c), we have the full list of temperatures to be used for our temperature intervals in the subsequent step (e).

Hot composite intervals (°C):   200—180—*140—120—*110—90—65—50

Cold composite intervals (°C)   180—160—*120—100—*90—70—45—30

(e) We now tabulate our temperature intervals and compute the temperature difference in each interval as shown below.

| S/N | Temperature intervals [°C] | | Temperature difference ΔT |
	Hot composite	Cold composite	
1	180–200	160–180	20
2	140–180	120–160	40
3	*140–140	*120–120	–
4	120–140	100–120	20
5	110–120	90–100	10
6	*110–110	*90–90	–
7	90–110	70–90	20
8	65–90	45–70	25
9	50–65	30–45	15

Special Treatment for Reboiler and Condenser

For our reboiler and condenser, the temperature interval has zero temperature difference since **temperature is fixed during phase change**. We cannot omit them from the stream table because of an absence of a temperature difference since these units still have a non-zero enthalpy load (duty of 3.2 MW) and have to be included in the energy cascade analysis.

(f) Next, we note the individual streams present in each interval.

- For example, if we have hot stream 3 that is cooled from 200 to 120 °C, we will indicate its stream label "3H" in the relevant temperature intervals that cover its temperature range from supply to target. The three intervals applicable for 3H are intervals 1, 2, and 4 as marked in red below.
- Similarly, we can do so for cold streams. For example, stream 1 is heated from 70 °C to 160 °C, so indicate the label 1C in intervals 2, 4, 5, and 7 as marked in blue below.

• Note that the condenser and reboiler units will contain only their own stream numbers.

We will obtain all inputs shown in the table below after completing part (f) for all the streams.

S/N	Temperature Intervals [°C]		Temperature difference ΔT	Streams present in interval
	Hot Composite	Cold Composite		
1	180-200	160-180	20	3H
2	140-180	120-160	40	3H, 1C, 2C
3	*140-140	*120-120	-	5C
4	120-140	100-120	20	3H, 1C, 2C
5	110-120	90-100	10	6H, 1C, 2C
6	*110-110	*90-90	-	4H
7	90-110	70-90	20	6H, 1C, 2C
8	65-90	45-70	25	6H, 2C
9	50-65	30-45	15	6H

(g) We next compute the total enthalpy change ΔH within each temperature interval.
   For process streams:

• $\Delta H = \sum_i W_i(\Delta T) = \sum W_C - W_H$
• We **sum contributions from all hot and cold streams present within that interval**. From the expression, we see that if the process streams cumulatively lose heat to the surroundings, then $\Delta H < 0$ and vice versa.
• The ΔH values for each temperature interval is as shown below.

  – For interval 1: $\Delta H = -W_3(20) = -90(20) = -1800$
  – For interval 2: $\Delta H = (-W_3 + W_1 + W_2)(40) = (-90 + 140 + 50)$
     $(40) = +4000$
  – For interval 4: $\Delta H = (-W_3 + W_1 + W_2)(20) = (-90 + 140 + 50)$
     $(20) = +2000$
  – For interval 5: $\Delta H = (-W_6 + W_1 + W_2)(10) = (-85 + 140 + 50)$
     $(10) = +1050$
  – For interval 7: $\Delta H = (-W_6 + W_1 + W_2)(20) = (-85 + 140 + 50)$
     $(20) = +2100$
  – For interval 8: $\Delta H = (-W_6 + W_2)(25) = (-85 + 50)(25) = -875$
  – For interval 9: $\Delta H = -W_6(15) = -85(15) = -1275$

  For reboiler and condenser:

• They operate at constant temperature, so there is no temperature change.
• These units are typically supplied at defined specifications (e.g., duties); therefore we simply use their stipulated duties as the enthalpy change values.

- In the problem, the enthalpy change for both is equivalent to 3.2 MW or 3200 kW.

Filling in the values computed above into our table, we have:

S/N	Temperature intervals [°C]		Interval size ΔT	Streams present	Enthalpy Change ΔH [kW]	
	Hot Composite	Cold Composite			Heating or cooling	Phase change
1	180-200	160-180	20	3H	-1800	-
2	140-180	120-160	40	3H, 1C, 2C	4000	-
3	*140-140	*120-120	-	5C	-	3200
4	120-140	100-120	20	3H, 1C, 2C	2000	-
5	110-120	90-100	10	6H, 1C, 2C	1050	-
6	*110-110	*90-90	-	4H	-	-3200
7	90-110	70-90	20	6H, 1C, 2C	2100	-
8	65-90	45-70	25	6H, 2C	-875	-
9	50-65	30-45	15	6H	-1275	-

We can tidy up our table by combining results for the last two columns marked with a bold border ("heating or cooling" and "phase change") into a single column as shown below.

S/N	Temperature intervals [°C]		Enthalpy change ΔH [kW] of process streams
	Hot composite	Cold composite	
1	180–200	160–180	−1800
2	140–180	120–160	4000
3	*140–140	*120–120	3200
4	120–140	100–120	2000
5	**110–120**	**90–100**	**1050**
6	*110–110	*90–90	−3200
7	90–110	70–90	2100
8	65–90	45–70	−875
9	50–65	30–45	−1275

It is important to be clear that when the process streams lose heat to their external **surroundings, the external surroundings gains or receives the same amount of heat**. In other words, $\sum \Delta H$ of process streams $= - \sum \Delta H$ of external surroundings. Or equivalently, **ΔH of process streams $= - \Delta H$ of external surroundings**. The enthalpy change of external surroundings can also be understood as the excess heat cascaded down (or excess heat rejected) by the process streams.

We have included a new column shown below, with the corresponding enthalpies for the external surroundings. They are of the same value as the enthalpies for process streams, but with the opposite sign.

S/N	Temperature intervals [°C]		Enthalpy change of process streams, ΔH [kW]	Enthalpy change of external surroundings, $-\Delta H$ [kW]
	Hot Composite	Cold Composite		
1	180-200	160-180	-1800	1800
2	140-180	120-160	4000	-4000
3	*140-140	*120-120	3200	-3200
4	120-140	100-120	2000	-2000
5	110-120	90-100	1050	-1050
6	*110-110	*90-90	-3200	3200
7	90-110	70-90	2100	-2100
8	65-90	45-70	-875	875
9	50-65	30-45	-1275	1275

(h) We will now construct our energy cascade diagram as shown below, where we present the temperature intervals from the table as a pair of vertical axes in descending order.

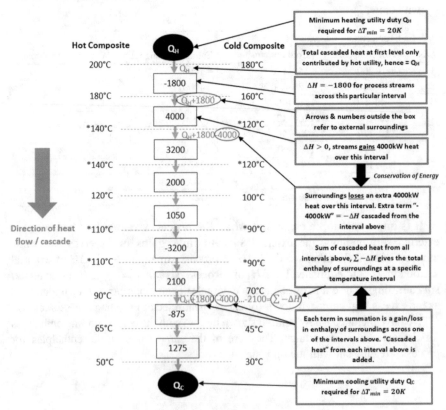

Here are some key points to note about the energy cascade diagram above:

- We add external heating and cooling utilities Q_H and Q_C at the top and bottom of the cascade respectively.

    - A heating utility supplies a non-zero enthalpy at the top (or start) of the cascade, serving as a heat source to the first interval.
    - A cooling utility receives all excess enthalpy at the bottom (or end) of the cascade, serving as a heat sink from the last interval.

- The **numbers in the boxes represent enthalpy change ΔH of process streams** within a specific temperature interval, i.e., "-1800" in the first box, "4000" in the second box, etc. going downwards.

    - Its value is the sum of all heating and cooling requirements within that particular interval which we have earlier computed under step (g). The hot streams in the interval expel heat, while the cold streams absorb heat.

- **Heat energy should balance at each interval**. The consequence of this is that **any excess heat from one interval is passed down to the next interval** at a lower temperature. This is represented by the arrows in orange and we may also call this "cascaded heat."
- The "**cascaded heat**" is shown by **arrows and values indicated outside the boxes and beside the arrows**. Since they are outside the boxes (boxes representing process streams), the cascaded enthalpies are representative of the **enthalpy change of the external surroundings, $-\Delta H$**.

    - The cascaded heat **cumulates or sums as they pass through intervals**, hence a growing sum of values added to Q_H as we proceed down the intervals, i.e., "$Q_H \rightarrow Q_H + 1800 \rightarrow Q_H + 1800 - 4000\ldots$"
    - The amount of cascaded heat out of each box is equivalent to $-\Delta H$ for that interval, and the total amount of heat cascaded to any particular interval is the sum of all cascaded heat from all the boxes before (or above) it, i.e., $\sum -\Delta H$.
    - The cascaded enthalpies at every interval have to be non-negative in the direction downwards, since heat cannot travel from cold to hot. This means the values "Q_H," "$Q_H + 1800$," "$Q_H + 1800 - 4000$" etc. have to be non-negative.

Using Cascaded Heat to Find Pinch Point

We can further **simplify the expressions for cumulative cascaded heat $\sum -\Delta H$ at each temperature interval** as shown in the table below and corresponding energy cascade diagram below (in bold orange). In the diagram, we have denoted the cascaded heat (or enthalpy) from the topmost external heating utility as Q_H. Similarly, the total cascaded heat received by the bottom-most external cooling utility is denoted Q_C.

S/N	Temperature intervals [°C]		Enthalpy change of process streams, ΔH [kW]	Enthalpy change of external surroundings, $-\Delta H$ [kW]	$\sum -\Delta H$ (i.e. sum of terms added to Q_H in cascade diagram)
	Hot Composite	Cold Composite			
1	180-200	160-180	-1800	1800	**1800**
2	140-180	120-160	4000	-4000	1800-4000 = **-2200**
3	*140-140	*120-120	3200	-3200	1800-4000-3200 = **-5400**
4	120-140	100-120	2000	-2000	**-7400**
5	**110**-120	**90**-100	1050	-1050	**-8450** (most negative)
6	*110-110	*90-90	-3200	3200	**-5250**
7	90-110	70-90	2100	-2100	**-7350**
8	65-90	45-70	-875	875	**-6475**
9	50-65	30-45	-1275	1275	**-5200**

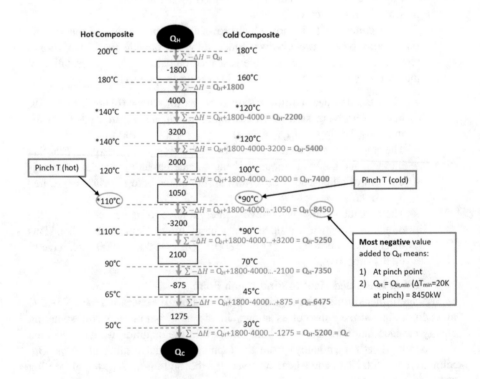

From the above diagram, the pinch point can be simply identified as the point where we have the **most negative value (−8450 kW) of cascaded heat** added to Q_H. This occurs at the temperature **110 °C (hot pinch temperature)** and **90 °C (cold pinch temperature)**, indicated in bold in the table above. Another result we can immediately tell is the value of Q_H in order to achieve $\Delta T_{min} = 20$ K at the pinch is simply **8450 kW**. This is also referred to as $Q_{H,min}$ since it is the minimum heating utility required for a system that has a condition of $\Delta T_{min} = 20$ K.

Why is the pinch point located where the most negative $\sum - \Delta H$ is added to Q_H?

Consider the case **if we had no external heating utility, i.e., $Q_H = 0$,** then we will encounter negative values of cascaded heat at one or more temperature intervals, implying that heat flows upwards from cold to hot which is not possible.

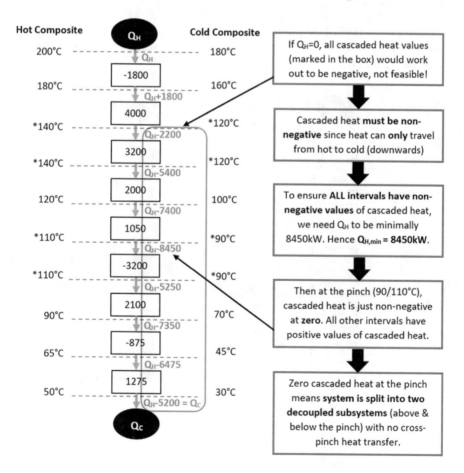

Finding the value of $Q_{H,min}$

To abide by the heat flow direction from hot to cold, **all values of "cascaded heat" must be non-negative pointing downwards**. In order to achieve this, we can **add $Q_H = 8450$ kW to all levels of the cascade** such that the most negative value of $\sum - \Delta H = -8450$ kW at interval 5 just becomes non-negative (i.e., zero).

Tracing back to the top of the cascade, this value of Q_H (or more specifically $Q_{H,min}$) is therefore also the hot utility duty that has to first cascade down from the top. [The word "min" in the subscript is used because we are fulfilling the minimum approach temperature ΔT_{min} condition. Refer to Problem 3 for more on this.]

Note also that we cannot simply add heat (Q_H in this case) at only one or a few of the intervals without doing so at **all** intervals by the **same amount** Q_H. This is because the added heat has to follow through all the way down the cascade to ensure energy conservation.

Finding the value of $Q_{C,min}$

As for the bottom of the cascade, we need to ensure energy balance by adding an external cooling utility Q_C which receives all residual heat cascaded out from the last temperature interval.

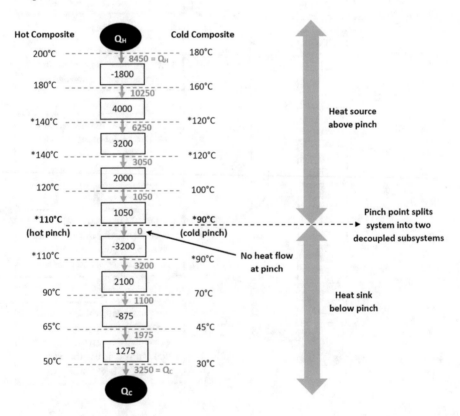

At the tail end of the cascade, we sum up all amounts of cascaded heat and arrive at $Q_{C,\,min} = 3250$ kW.

$$Q_{C,min} = Q_{H,min} - 5200 = 8450 - 5200$$

$$Q_{C,min} = 3250 \text{ kW}$$

Now we may finally update our energy cascade diagram with the computed values of $Q_{H,\,min}$ and $Q_{C,\,min}$ as well as all values of cascaded heat between intervals. The pinch temperatures **110 °C (hot side) and 90 °C (cold side)** are also indicated in the diagram.

Our results are summarized as shown.

$$T_{pinch,hot} = 110^{\circ}\text{C}, \quad T_{pinch,cold} = 90^{\circ}\text{C}, \quad Q_{H,min} = 8450 \text{ kW}, \quad Q_{C,min} = 3250 \text{ kW}$$

(c)

We notice that there is **zero cascaded heat (or no flow of energy) across the pinch point,** or in other words cross-pinch heat transfer should be zero in the ideal scenario. The significance of the pinch is therefore in separating our system into two distinct parts; the top half requires a heat input and is therefore a **net heat sink (i.e., heat deficit above pinch)**, while the bottom half rejects heat and is therefore a **net heat source (i.e., heat surplus below pinch)**. These two parts are also referred to as "**decoupled subsystems**" whereby there is no heat flow from one subsystem to the other.

Any **heat leakage** from the hotter region above the pinch (heat sink) to the colder region below the pinch (heat source) **results in an increase in BOTH the external heating and cooling utilities required** since the heating utility Q_H needs to increase by the same amount of heat that is lost to the portion below the pinch, while the cooling utility Q_C needs to similarly increase by the amount required to accept the extra "undesired" heat gained from the top portion above the pinch through cross-pinch leakage.

This is sometimes referred to as a "double penalty" due to the increased energy requirement on both ends of the cascade. We can therefore appreciate why engineers try to minimize cross-pinch heat transfer as it introduces inefficiencies and works against heat recovery objectives.

Problem 5

Heat integration is required for the following streams in a process plant.

Stream label	Flow rate heat capacity [kW/K]	Supply temperature [°C]	Target temperature [°C]
A	14	210	50
B	6	180	90

(continued)

Stream label	Flow rate heat capacity [kW/K]	Supply temperature [°C]	Target temperature [°C]
C	12	40	170
D	22	100	150

Assuming a minimum approach temperature of $\Delta T_{min} = 10$ K, and using shifted temperature intervals,

(a) **Determine the pinch temperatures and minimum external utility duties required.**
(b) **Using your results above, draw a grid display for this system showing the design of a heat exchanger network that achieves maximum energy recovery.**

Solution 5

(a)

We can make use of a simple "trick" to find minimum utilities and the pinch point. This method uses the so-called **shifted (or adjusted) temperatures** instead of original stream temperatures (as seen in Problem 4) as they help "transform" our original problem into a simpler one to solve. After obtaining your final answers using shifted temperatures, we can then simply "undo" the shift (or adjustment) to arrive at the actual (pre-shift) stream temperatures.

To find shifted temperatures, we need to make use of the given minimum approach temperature $\Delta T_{min} = 10$ K:

- **Add** $\frac{1}{2}\Delta T_{min}$ to all **cold** stream temperatures (i.e., cold composite curve shifts up by $\frac{1}{2}\Delta T_{min}$)
- **Subtract** $\frac{1}{2}\Delta T_{min}$ from all **hot** stream temperatures (i.e., hot composite curve shifts down by $\frac{1}{2}\Delta T_{min}$)

Visually we can observe the effect of this shift on the hot and cold composite curves as follows. We see that the shift helps transform the pinch point from having a separation of 10 K between the hot and cold composites to having zero separation, i.e., hot and cold stream lines at shifted temperatures exactly **meet at the pinch**. The usefulness of this is to avoid having to deal with two sets of temperatures (hot and cold composite temperatures like in Problem 4) when locating the pinch point, and instead to only deal with one set of temperatures (shifted). This works because for **both** the hot and cold composite lines, shifted temperatures at the pinch are equivalent, hence fulfilling $\Delta T_{min}(\text{shifted}) = 0$.

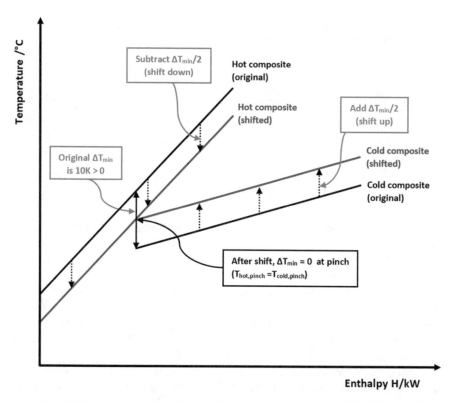

Returning to our problem, we can therefore determine shifted temperatures, denoted T^* as shown below. Note that we have also labeled the streams either hot or cold by observing whether a stream's target temperature is higher or lower than its supply temperature.

Stream label	Heat capacity, W [kW/K]	Supply temperature [°C]	Target temperature [°C]	Supply temperature (shifted), T_S^* [°C]	Target temperature (shifted), T_T^* [°C]
A (hot)	14	210	50	$210 - \frac{1}{2}(10)$ $= 210 - 5 = \mathbf{205}$	$50 - \frac{1}{2}(10)$ $= 50 - 5 = \mathbf{45}$
B (hot)	6	180	90	$180 - 5 = \mathbf{175}$	$90 - 5 = \mathbf{85}$
C (cold)	12	40	170	$40 + \frac{1}{2}(10)$ $= 40 + 5 = \mathbf{45}$	$170 + \frac{1}{2}(10)$ $= 170 + 5 = \mathbf{175}$
D (cold)	22	100	150	$100 + 5 = \mathbf{105}$	$150 + 5 = \mathbf{155}$

We can now tabulate a series of enthalpy values using shifted temperature intervals. First, we identify the **streams present in each interval**, then we **compute the net heat capacities for each interval** (refer to Problem 4 part (b), step (g) for details) and compute the **corresponding enthalpies** and **cumulative enthalpies** at each interval. The results are shown below.

T^* interval [°C]	Streams present	Net heat capacity, $\sum W_C - W_H$ [kW/K]	ΔH of process streams [kW]	$-\Delta H$ [kW]	$\sum - \Delta H$ [kW]
205–175	A	−14	−14 (30) = −420	420	420
175–155	A, B, C	−14 − 6 + 12 = −8	−160	160	420 + 160 = 580
155–105	A, B, C, D	14	700	−700	−120 (most negative)
105–85	A, B, C	−8	−160	160	40
85–45	A, C	−2	−80	80	120

The pinch point occurs where $\sum - \Delta H$ is the most negative, and the minimum heating utility required $Q_{H,\,min}$ for $\Delta T_{min} = 10$ K at pinch is also exactly equivalent to the absolute value of this most negative number, i.e., $Q_{H,\,min} = 120$ kW (refer to Problem 4, part (b) for details).

We may construct an energy cascade diagram from the tabulation above, this time using shifted temperatures for our cascade (unlike in Problem 4). By using shifted temperatures, we will only need to refer to a single set of temperatures on the temperature axis of the cascade, instead of two separate axes (one each for the hot and cold composite—refer to Problem 4 for details).

From the cascade, we note that the pinch occurs when $T^* = 105\ °C$ **since this is the point where no heat flows downwards.** To find the actual hot and cold stream temperatures at this pinch, we simply add 5 °C back to this value (i.e., 105 °C) for the hot pinch temperature and subtract 5 °C from 105 °C for the cold pinch temperature.

The results obtained so far are summarized below:

$$T_{pinch,hot} = 110\,°C, \quad T_{pinch,cold} = 100\,°C, \quad Q_{H,\,min} = 120\ kW$$

We can deduce the value of $Q_{C,\,min}$ once $Q_{H,\,min}$ is known and compute its value by following through the cascade until the bottom end as shown below.

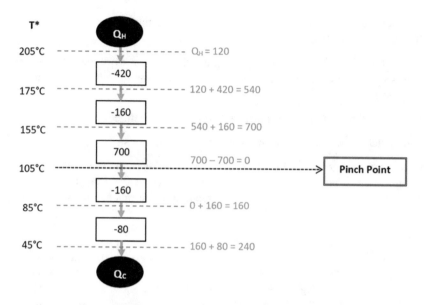

We obtain the minimum cooling utility $Q_{C,min} = 240$ kW.

(b)

The grid display for maximum energy recovery involves matching hot streams to cold streams as much as possible to achieve maximum heat integration. Any remaining heating or cooling loads that cannot be fulfilled by matches may then be met by the adding of external utilities.

There are some conditions for HEN design to take note of, and they are shown as follows:

Above the Pinch (Hot Side)

- Number of cold streams should be greater or equal to the number of hot streams, $N_C \geq N_H$.
- Only heating utilities should be present, Q_H.

Below the Pinch (Cold Side)

- Number of hot streams should be greater or equal to the number of cold streams, $N_H \geq N_C$.
- Only cooling utilities should be present, Q_C.

Matches at the Pinch (i.e., if stream temperatures = pinch temperatures)

- On the hot side of the pinch, heat capacities of matching streams must fulfill $W_C \geq W_H$.
- On the cold side of the pinch, heat capacities of matching streams must fulfill $W_H \geq W_C$.

It also helps to be cognizant of certain tips or "rules-of-thumb" that would help us match streams easily; they are:

- **Start at the pinch**, then continue placing matches as you work away the pinch towards the hot end (section above pinch) and towards the cold end (section below pinch).

  - Check that **conditions for "matches" at pinch** are fulfilled.

- Select matches that are best at satisfying **entire load(s)** of streams.

When drawing a grid display, line up all process streams as shown below, indicating the pinch temperatures, the respective heat capacities (denoted FRHC), and heating or cooling loads of the streams relative to the pinch. This is done as shown below.

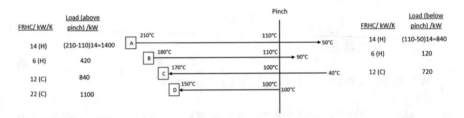

From the above diagram, we observe that above the pinch on the left-hand side, $N_H = 2$ (streams A and B) and $N_C = 2$ (streams C and D); therefore $N_C \geq N_H$ is fulfilled. Below the pinch on the right-hand side, $N_H = 2$ (streams A and B) and $N_C = 1$ (stream C); therefore $N_H \geq N_C$ is also fulfilled.

Intermediate Calculations for Stream Matches

Whenever there is a match between streams, there is essentially heat transfer between the coupled streams via a heat exchanger unit (HXer). Upon exiting a HXer representing a stream match, it is good practice to calculate intermediate temperatures of the streams (i.e., temperatures of the streams when they leave the HXer), residual loads of the streams (if target temperatures not reached yet), and HXer duties. These are further explained below.

- **Intermediate temperatures**

  - Temperatures of process streams **after they exit the HXer for the stream match**.
  - Hot stream would have cooled down while the cold stream would have heated up as they pass through the HXer.
  - The **same amount of heat lost by the hot stream is gained by the cold stream** in the match, and this amount of heat energy exchanged defines the HXer duty for the stream match.

- **Residual loads**

  - After exiting a stream match, the hot stream may still have a residual cooling load if the **match did not completely fulfill its cooling load**. In other words, the hot stream has not reached its defined target temperature although it will be closer to it than before the match. Conversely, the cold stream may have a residual heating load after the match.
  - In a match between two streams of unequal loads, the **stream with the smaller load will be completely fulfilled** and the stream with the larger load will be left with a residual load equivalent to the difference between the two loads.
  - Any residual loads should be computed to decide if we need to **place more matches and/or add external utilities** to fulfill all residual loads.

- **HXer duty**

  - The HXer representing a stream match brings a hot stream and cold stream in close proximity for inter-stream heat transfer. The amount of heat transferred between the streams within the HXer unit is also the HXer duty in kW.

Matches at the Pinch

On the **hot side of the pinch**, we must check that $W_C \geq W_H$ **at the pinch**. We notice that cold stream C ($W_C = 12$) can only match with hot stream B ($W_H = 6$) but not with hot stream A ($W_H = 14$). Therefore, it makes sense to "reserve" stream B for stream C, while the remaining stream A can match with cold stream D and still fulfill $W_C \geq W_H$.

In the same way, we can place matches on the **cold side of the pinch**, but this time choosing only streams whereby $W_H \geq W_C$. In this case, cold stream C can only match with hot stream A but not with hot stream B. Therefore, the only logical match is between streams A and C at the pinch.

To summarize, we have the following three matches at the pinch:

1. Streams A and C (cold side of pinch)
2. Streams B and C (hot side of pinch)
3. Streams A and D (hot side of pinch)

Match 1: Between A and C (Cold Side)

Match 1 is drawn in our grid diagram as shown below in red with the corresponding residual load of 120 kW and intermediate temperature 59 °C of stream A indicated. [Stream C's load below the pinch is completely fulfilled by this match; hence intermediate temperature and residual load calculations are not required.]

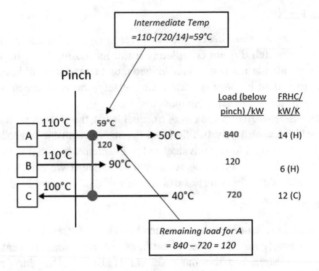

Intermediate Temperatures

Stream A enters Match 1 with a temperature of 110 °C (hot pinch). Within the HXer representing Match 1, it loses 720 kW of heat to stream C. We can calculate the temperature difference ($=720/14$) of stream A between the inlet and outlet of the HXer using its heat capacity value of 14 kW/K. Hence the exit temperature (or intermediate temperature) of stream A is $110 - (720/14) = 58.6$ or approximately 59 °C as shown in the diagram.

Residual Loads

In Match 1, stream C's load is fully fulfilled since its 720 kW heating load required to raise its supply temperature at 40 °C to the cold pinch temperature at 100 °C can be wholly provided by stream A. And since stream A has a total load of 840 kW, it will be left with a residual load of 120 kW, after supplying 720 kW to C.

A stock-take of unfulfilled loads after Match 1 is placed is (1) a full load for B at 120 kW and (2) a residual load of 120 kW for A. We had found earlier that the minimum external cooling utility required for our system was $Q_{C,min} = 240$ kW; this means that we can fulfill the total residual loads using external utilities and need not place any further matches. Note that the cold utility can be split into two parts, such that the total **240 kW minimum utility duty is supplied via two separate 120 kW cold utilities added to hot streams A and B respectively**.

We have now completed our HEN design for the section **below the pinch** as shown in the grid display below, where all three streams' loads are fulfilled.

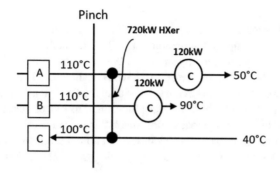

HXer Duty

The HXer duty for Match 1 is 720 kW, since it delivers 720 kW of heat from hot stream A to cold stream C.

Match 2: Between B and C (Hot Side)

Following the same steps as Match 1, we may now work on the section above the pinch (hot side) and place Matches 2 and 3 there, starting at the pinch.

Match 2 is drawn in our grid diagram as shown below in purple with the corresponding residual load of 420 kW and intermediate temperature 135 °C of stream C indicated. [Stream B's load above the pinch is completely fulfilled by this match.] The HXer duty of 420 kW for this match is also indicated, whereby 420 kW is the amount of heat transferred from hot stream B to cold stream C in the HXer.

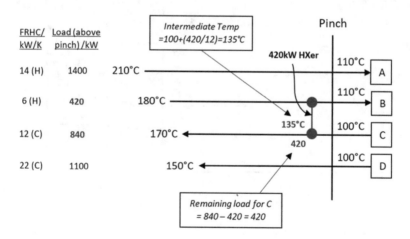

Intermediate Temperatures

Stream C enters Match 2 with a temperature of 100 °C (cold pinch). Within the HXer representing Match 2, stream C gains 420 kW of heat from stream B. We can calculate the temperature difference (=420/12) of stream C between the inlet and outlet of the HXer using its heat capacity value of 12 kW/K. Hence the exit temperature (or intermediate temperature) of stream C is 100 + (420/12) = 135 °C as shown in the diagram.

Residual Loads

In Match 2, stream B's load is fully fulfilled since its 420 kW cooling load required to lower its supply temperature at 180 °C to the hot pinch temperature at 110 °C can be wholly provided by stream C. We can confirm that stream B's load is fulfilled by using the heat capacity of B (6 kW/K) as shown: 6(180 − 110) = 420 kW (confirmed). Separately, since stream C has a total load of 840 kW, it will be left with a residual heating load of 420 kW, after absorbing 420 kW of heat from B.

A stock-take of unfulfilled loads after Match 2 is placed is (1) a full load for A at 1400 kW, (2) a residual load of 420 kW for C, and (3) a full load of 1100 kW for D. We will address these residual loads by placing the next match at the pinch, i.e., Match 3.

Match 3: Between A and D (Hot Side)

Another match that can be done at the pinch (on the hot side) is between A and D.

Match 3 is now added to our grid diagram as shown below in purple with the corresponding residual load of 300 kW and intermediate temperature 189 °C of stream A indicated. [Stream D's load above the pinch is completely fulfilled by this match.] The HXer duty of 1100 kW for this match is also indicated, whereby 1100 kW is the amount of heat transferred from hot stream A to cold stream D in the HXer.

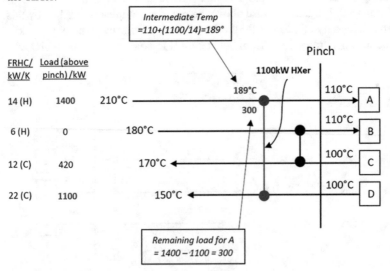

Intermediate Temperatures

Stream A has to exit Match 3 with a temperature of 110 °C (hot pinch). Within the HXer representing Match 3, 1100 kW of heat is lost from A to D. We can calculate the temperature difference (=1100/14) of stream A between the inlet and outlet of the HXer using its heat capacity value of 14 kW/K. Hence the intermediate temperature of stream A at the inlet of the HXer can be back-calculated as 110 + (1100/ 14) = 188.6 or approximately 189 °C as shown in the diagram.

Residual Loads

In Match 3, stream D's load is fully fulfilled since its 1100 kW heating load required to raise its cold pinch temperature at 100 °C to its target temperature at 150 °C can be wholly provided by stream A. We can confirm that stream D's load is fulfilled by using the heat capacity of D (22 kW/K) as shown: 22 (150 − 100) = 1100 kW (confirmed). Separately, since stream A has a total load of 1400 kW, it will be left with a residual load of 300 kW, after supplying 1100 kW to D.

A stock-take of unfulfilled loads (on the hot side of the pinch) after Matches 2 and 3 are placed is (1) a residual load for A at 300 kW and (2) a residual load of 420 kW for C. Recall that the section above the pinch can only have hot utilities added (no cold utilities allowed); therefore the residual load of hot stream A still needs to be addressed by placing more stream matches.

We now move on to placing matches away from the pinch, while still working on the section above the pinch.

Matches Away from Pinch (Hot Side)
Match 4: Between A and C (Hot Side)

Since all streams have now moved away from the pinch temperatures, we need not fulfill the condition at the pinch $W_C \geq W_H$ on the hot side, for the next match. Since only hot stream A and cold stream C are left, it is logical to match them next.

The fourth Match (let us call it Match 4) between A and C is now added to our grid diagram as shown below in purple with the corresponding residual load of 120 kW and intermediate temperature 160 °C of stream C indicated. [Stream A's load above the pinch is now completely fulfilled by this match.] The HXer duty of 300 kW for this match is also indicated, whereby 300 kW is the amount of heat transferred from hot stream A to cold stream C in the HXer.

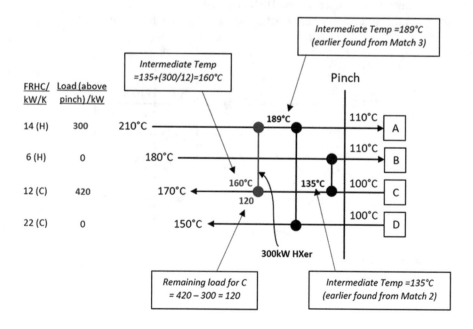

Intermediate Temperatures

Stream C enters Match 4 with a temperature of 135 °C which was the exit temperature from Match 2's HXer (earlier calculated).

Within the HXer representing Match 4, 300 kW of heat is lost from A to C. We can calculate the temperature difference (=300/12) of stream C between the inlet and outlet of the HXer using its heat capacity value of 12 kW/K. Hence the exit temperature (or intermediate temperature) of stream C is 135 + (300/12) = 160 °C as shown in the diagram.

Residual Loads

In Match 4, stream A's residual 300 kW cooling load is fully fulfilled since it is able to cool down from its supply temperature at 210 °C to the intermediate temperature of 189 °C (before going into Match 3) by losing 300 kW of heat to stream C. We can confirm this using the heat capacity of A (14 kW/K) as shown: 14 (210 − 189) = 300 kW (confirmed). Separately, since stream C had a residual heating load of 420 kW prior to Match 4, it will be left with a residual load of 120 kW after Match 4, as it gains 300 kW of heat from stream A from this match.

A stock-take of unfulfilled load(s) after all matches up to Match 4 is only a residual load of 120 kW for C. We had found earlier from part (a) that the minimum external heating utility was $Q_{H,min}$ = 120 kW; this means that we can fully satisfy the residual load on C by adding an external hot utility of 120 kW duty and we need not place further matches. We can do a quick calculation to confirm that the 120 kW external heating utility will ensure that stream C reaches its target temperature of 170 °C after exiting Match 4 (HXer of 300 kW) with an intermediate temperature of 160 °C. Using the heat capacity of C (12 kW/K), we have 12(170 − 160) = 120 kW (confirmed).

We have now completed our HEN design for the section above the pinch as shown in the grid display below, where all four streams' loads are fulfilled.

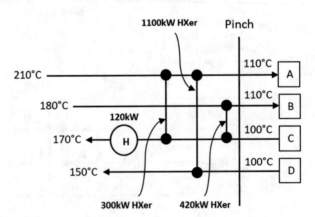

Combining our results for the sections above and below the pinch, we arrive at our final HEN design for maximum energy recovery (MER) at ΔT_{min} = 10 K for the entire system. There is a total of four HXers, two cold utilities, and one hot utility.

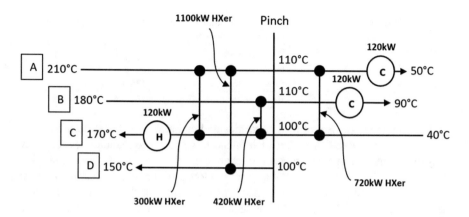

Problem 6

Consider four streams below with the following specifications.

Stream	Flow rate heat capacity [kW/K]	Supply temperature [°C]	Target temperature [°C]
1	6	160	50
2	3	140	20
3	4	10	125
4	8	70	130

Using a minimum approach temperature of 10 K,

(a) **Determine the pinch temperature and required external utility duties and show your results in an energy cascade diagram.**

(b) **Construct the hot and cold composite curves for this system and highlight any key features.**

(c) **Draw a grid display showing the heat exchanger network that achieves maximum energy recovery.**

Solution 6

(a)

Like in Problem 5, we can start by tabulating our shifted temperatures and identifying which streams are hot (i.e., supply temperature > target temperature) and which are cold (i.e., supply temperature < target temperature).

Given a minimum approach temperature $\Delta T_{min} = 10$ K, we can find shifted temperatures by following the steps here:

- Add $\frac{1}{2}\Delta T_{min} = 5$ K to all cold stream temperatures
- Subtract $\frac{1}{2}\Delta T_{min} = 5$ K from all hot stream temperatures

Stream	Flow rate heat capacity [kW/K]	Supply temperature (shifted), T_S^* [°C]	Target temperature (shifted), T_T^* [°C]
1 (H)	6	160 − 5 = **155**	50 − 5 = **45**
2 (H)	3	140 − 5 = **135**	20 − 5 = **15**
3 (C)	4	10 + 5 = **15**	125 + 5 = **130**
4 (C)	8	70 + 5 = **75**	130 + 5 = **135**

We now tabulate a set of heat capacities and enthalpies for the **shifted temperature intervals**. Note that within each interval, we compute the total heat capacity by considering all streams present in that interval according to the expression $\sum W_C - W_H$. Since cold streams gain heat (endothermic) and hot streams lose heat (exothermic), cold streams are assigned a positive sign for the heat capacities and hence enthalpies, and conversely hot streams are assigned corresponding negative signs.

- Net flow rate heat capacity (kW/K) within a temperature interval, $\sum W_C - W_H$
- If $\sum W_C - W_H > 0$ in a temperature interval, then $\Delta H > 0$, which implies that all streams in the interval sum up to a net heat gain. Conversely if $\Delta H < 0$ then all streams in the temperature interval sum up to a net heat loss.

The streams present within each shifted temperature interval are tabulated below. For each interval, we calculate its corresponding net flow rate heat capacity, $\sum W_C - W_H$, and enthalpy change of the streams, ΔH. To construct an energy cascade diagram and identify the pinch point easily, we have also added columns for $-\Delta H$ (negative of enthalpy change of process streams or enthalpy change of external surroundings) and $\sum -\Delta H$ or the cumulative enthalpy change of external surroundings at a particular interval.

T^* interval [°C]	ΔT [°C]	Streams present	$\sum W_C - W_H$ [kW/K]	ΔH of process streams [kW]	$-\Delta H$ [kW]	$\sum -\Delta H$ [kW]
155–135	20	1	−6	−6 (20) = −120	120	120
135–130	5	1, 2, 4	−6 − 3 + 8 = −1	−1(5) = −5	5	120 + 5 = 125
130–75	55	1, 2, 3, 4	−6 − 3 + 4 + 8 = 3	3(55) = 165	−165	120 + 5−165 = −40 (most negative)
75–45	30	1, 2, 3	−6 − 3 + 4 = −5	−5 (30) = −150	150	120 + 5 − 165 + 150 = 110
45–15	30	2, 3	−3 + 4 = 1	1(30) = 30	−30	120 + 5 − 165 + 150 − 30 = 80

The pinch point occurs at the interval where $\sum -\Delta H = -40$ since this is the **most negative value of** $\sum -\Delta H$ in the column. We can directly determine the pinch temperature (shifted) to be $T_{pinch}^* = 75\,°C$ (or in terms of actual temperatures, $T_{pinch, hot} = 75 + 5 = 80\,°C$, $T_{pinch,cold} = 75 - 5 = 70\,°C$). The minimum heating utility can also be immediately determined, as it is simply the value of this most negative $\sum -\Delta H$, i.e., $Q_{H,min} = 40$ **kW**. To find the minimum cooling utility, $Q_{C,\,min}$, we need to construct an energy cascade and compute the series of cumulative enthalpies until we reach the bottom where we obtain the value of $Q_{C,min} = 120$ **kW**.

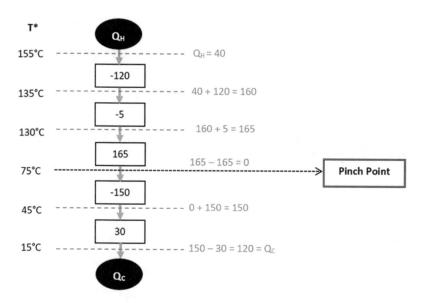

(b)

The composite curve refers to a plot of **temperature against enthalpy** of process streams. We are required to draw two curves, a hot composite curve (sum of hot streams present in a particular temperature interval) and a cold composite curve (sum of cold streams present in a particular temperature interval).

Step 1: Identifying streams present within each interval

We note from earlier that streams 1 and 2 are hot streams while 3 and 4 are cold streams. Given the supply and target temperatures of each stream, we can identify the hot streams present within each temperature interval and do the same for the cold streams. [Note: **Actual stream temperatures** (not shifted) are used to plot the composite curves here.]

Stream	Supply Temperature [°C]	Target Temperature[°C]
1 (H)	160	50
2 (H)	140	20
3 (C)	10	125
4 (C)	70	130

Hot Composite Curve

Temp intervals [°C]	ΔT [°C]	Hot streams present
20-50	30	2
50-140	90	1, 2
140-160	20	1

Cold Composite Curve

Temp intervals [°C]	ΔT [°C]	Cold streams present
10-70	60	3
70-125	55	3, 4
125-130	5	4

Step 2: Finding total heat capacity $\sum W_i$ and enthalpy change ΔH for each interval

For each composite curve, we can compute the total (or composite) heat capacity for each interval, i.e., $\sum_i W_{\mathrm{H},i}$ for the hot composite curve and $\sum_i W_{\mathrm{C},i}$ for the cold composite curve. This is done by summing heat capacities of all streams present within an interval.

The corresponding enthalpy changes ΔH [kW] for each interval can also be determined using the formula,

$$\Delta H = \Delta T \sum W_i$$

where W_i [kW/K] denotes flow rate heat capacity of one stream present in the interval. The computed values are shown in the tabulations below for the hot and cold composites, respectively.

Hot Composite Curve

Temp intervals [°C]	ΔT [°C]	Hot streams present	$\sum W_{H,i}$ [kW/K]	ΔH [kW]
20–50	30	2	3^i	$-30(3) = -90$
50–140	90	1, 2	$6 + 3 = 9$	$-90(9) = -810$
140–160	20	1	6	$-20(6) = -120$

We notice that $\Delta H < 0$ for hot composite curve since hot streams cool down, $\Delta T < 0$, and given that $\Delta H = \Delta T \sum W_i$, we obtain $\Delta H < 0$.

Cold Composite Curve

Temp intervals [°C]	ΔT [°C]	Cold streams present	$\sum W_{C,i}$ [kW/K]	ΔH [kW]
10–70	60	3	4^i	$60(4) = 240$
70–125	55	3, 4	$4 + 8 = 12$	$55(12) = 660$
125–130	5	4	8	$5(8) = 40$

We notice that $\Delta H > 0$ for cold composite curve since cold streams heat up, $\Delta T > 0$, and given that $\Delta H = \Delta T \sum W_i$, we obtain $\Delta H > 0$.

Step 3: Finding the gradient of composite curves

The hot and cold composite curves are each plotted by joining together consecutive line segments, whereby each line segment represents a temperature interval within which a particular number of streams are present. Since the composite curves are plotted on a vertical temperature axis and a horizontal enthalpy axis, the gradient of each line segment is also equivalent to the reciprocal of the total heat capacity of that interval.

$$\Delta H = \Delta T \sum W_i$$

$$\text{Gradient} = \frac{\text{rise}}{\text{run}} = \frac{\Delta T}{\Delta H} = \frac{1}{\sum W_i}$$

Using the correlation above, we can compute the gradients for each line segment as shown in the new column added to the tables below.

Hot Composite Curve

Temp intervals [°C]	ΔT [°C]	Hot streams present	$\sum_i W_{H,i}$ [kW/K]	ΔH [kW]	Gradient
20–50	30	2	3	$-30(3) = -90$	$\frac{1}{3} = 0.33$
50–140	90	1, 2	$6 + 3 = 9$	$-90(9) = -810$	$\frac{1}{9} = 0.11$
140–160	20	1	6	$-20(6) = -120$	$\frac{1}{6} = 0.17$

We notice that all gradient values for the line segments making up the hot composite curve are positive. Since $\Delta T < 0$ and $\Delta H < 0$, therefore $\frac{\Delta T}{\Delta H} > 0$.

Cold Composite Curve

Temp intervals [°C]	ΔT [°C]	Cold streams present	$\sum_i W_{C,i}$ [kW/K]	ΔH [kW]	Gradient
10–70	60	3	4	$60(4) = 240$	$\frac{1}{4} = 0.25$
70–125	55	3, 4	$4 + 8 = 12$	55 $(12) = 660$	$\frac{1}{12} = 0.083$
125–130	5	4	8	$5(8) = 40$	$\frac{1}{8} = 0.125$

We notice that all gradient values for the line segments making up the cold composite curve are also positive. Since $\Delta T > 0$ and $\Delta H > 0$, therefore $\frac{\Delta T}{\Delta H} > 0$.

Now that we have tabulated data for the gradient, temperature change, and enthalpy change for each temperature interval for both the hot and cold composite curves, we can plot the graph as shown below.

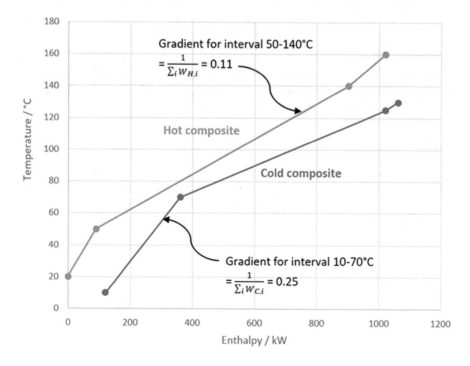

Key Features of the Composite Curves:

Direction of the Composite Curves

Since hot streams cool down as they go from supply to target temperatures, the hot composite curve starts from high temperature and enthalpy values at the top right corner of the plot and moves towards low temperature and enthalpy values at the bottom left corner of the plot. Conversely, since cold streams heat up from their supply to target temperatures, the cold composite curve runs from bottom left to top right, as it reaches towards higher temperature and enthalpy values.

Positions of the Composite Curves

The hot composite curve is always above the cold composite curve at all times and the two curves **cannot cross**. This ensures that heat always travels from a hotter to colder region (or hot streams to cold streams).

Step 4: Finding the pinch point and minimum utility duties from the plot

Pinch Point

From the plot, we can identify a point whereby the two composite curves are closest to each other (10 K apart) but not touching. This is known as the **minimum approach temperature,** $\Delta T_{min} = 10$ K where the pinch point is.

Note that the composite curves are closest to each other at the pinch but should not touch in all practical cases. If the curves touch, then $\Delta T_{min} = 0$ which means an infinitely large heat exchange area, which is not achievable in reality.

- Recall the design equation of a heat exchanger (HXer): $Q = UA\Delta T$, where Q [J/h] is the rate of heat transfer between two process streams passing through the HXer, U $\left[\frac{J}{h \cdot m^2 \cdot K}\right]$ represents the heat transfer coefficient while A [m^2] represents heat transfer area.
- If $\Delta T \rightarrow 0, \quad A \rightarrow \infty$.

The pinch point separates the system into two thermally isolated subsystems sections, above the pinch (hot end) and below the pinch (cold end). There is no heat transfer across the pinch.

Minimum Utility Duties

Above the pinch, the hot composite curve "overshoots" the cold composite curve lying below it by an amount equivalent to the minimum heating utility $Q_{H, min}$. Similarly, at the cold end, the cold composite curve "overshoots" the hot composite curve lying above it by $Q_{C, min}$, the minimum cooling utility.

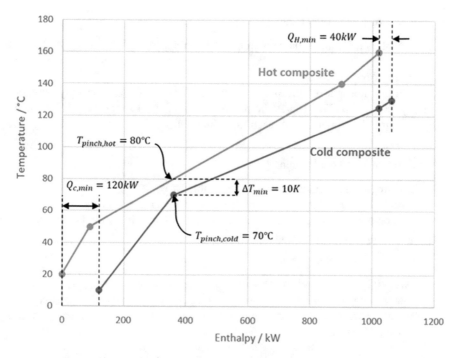

This length of "overshoot" is shown using dotted lines in the plot above, with the utility duties of $Q_{H, min} = 40$ kW, $Q_{C, min} = 120$ kW indicated.

(c)

Using the pinch temperatures earlier found, as well as the given supply and target temperatures, we can begin our grid display construction, starting with its basic structure as shown below. The respective cooling and heating loads of the streams, for the segments above and below the pinch, may be determined using flow rate heat capacity values (denoted FRHC) as follows (calculations shown for stream 1).

FRHC/ kW/K	Load (above pinch) /kW						FRHC/ kW/K	Load (below pinch) /kW
				Pinch				
6 (H)	(160-80)6=480	160°C ─ 1		80°C	→ 50°C		6 (H)	(80-50)6=180
3 (H)	180	140°C 2		80°C	→ 20°C		3 (H)	180
4 (C)	220		125°C 3 ←	70°C	10°C ← 4 (C)		4 (C)	240
8 (C)	480		130°C 4 ←	70°C				

Matters at the Pinch

Matches at the Pinch

At the Pinch (Cold Side)

Starting from the pinch point and working on the section below the pinch (or cold side), we can place our first match by selecting pairs of streams obeying $W_H \geq W_C$. This gives us only one possible match:

- Hot stream 1 and cold stream 3

  - Hot stream 2 cannot match with cold stream 3 since it violates $W_H \geq W_C$

At the Pinch (Hot Side)

As for the hot side of the pinch (above the pinch), we choose only streams whereby $W_C \geq W_H$. This leaves the remaining unmatched streams 2 and 3 which we are then able to match with each other since this pair at the pinch also obeys $W_C \geq W_H$.

- Hot stream 1 and cold stream 4

  - Stream 1 cannot match with 3 since it violates $W_C \geq W_H$.
  - Recall that one of the tips for matching streams is to satisfy the full loads of one or both streams as much as possible. This match fulfills both streams' loads completely so it is also a good match.

- Hot stream 2 and cold stream 3

  - Although stream 2 can match with both 3 and 4 and obey $W_C \geq W_H$, we have matched it with 3. This is because stream 1 is only able to match with stream 4, therefore we should reserve stream 4 for stream 1 and match stream 3 with stream 2 instead.

Considering all the matches at the pinch mentioned above, we arrive at the following grid diagram with intermediate temperatures indicated. [Refer to Problem 5 for details on calculating intermediate temperatures.]

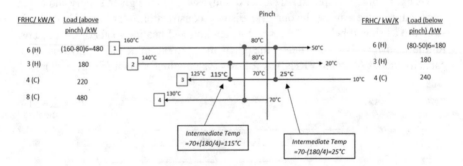

The HXer duties representing the three matches at the pinch are as follows:

- Match between 1 and 3 (below pinch)—180 kW
- Match between 1 and 4 (above pinch)—480 kW
- Match between 2 and 3 (above pinch)—180 kW

We can now compute any remaining loads that have not been fulfilled, to see if we require more stream matches and/or addition of external utilities to fulfill these residual loads completely.

Placing Matches Away from the Pinch

Below the Pinch

The match between streams 1 and 3 completely fulfills stream 1's load, while only partially fulfilling the heating load for stream 3. Heat from stream 1 helps to raise the temperature of cold stream 3 from 25 to 70 °C. Therefore, the remaining heating load for stream 3 that still needs to be fulfilled is equivalent to $4(25 - 10) = 60$ kW.

At the same time, we note that stream 2 still needs to be cooled from 80 °C at the pinch to 20 °C (its target temperature), which is equivalent to a cooling load of 180 kW.

Therefore, it makes sense for us to next match hot stream 2 with cold stream 3, such that stream 3's remaining load of 60 kW can be completely met by stream 2, and stream 3 will be left with a remaining cooling load that is equivalent to $180 - 60 = 120$ kW.

Since 120 kW is also the value of the minimum external cooling utility $Q_{C, min} = 120$ kW earlier found, we can place a cold utility on stream 2 to completely satisfy its residual load.

Above the Pinch

Above the pinch, streams 1 and 4's loads are both already satisfied from the earlier match between them at the pinch, since their loads are exactly equivalent at 480 kW.

The earlier match between streams 2 and 3 also completely fulfills stream 2's load. As for cold stream 3, it has a temperature of 70 °C at the pinch and heats up to a temperature of 115 °C after the match with stream 2. Therefore, stream 3 has a remaining amount of heating required from 115 °C to its final target temperature of 125 °C, which translates into a residual heating load of $4(125 - 115) = 40$ kW.

We found from part (a) that our minimum heating utility, $Q_{H, min} = 40$ kW. Since, therefore, the only remaining load left above the pinch is the heating load of 40 kW for stream 3, this can be exactly met by adding an external hot utility of 40 kW on stream 3.

We have therefore completed our HEN design for maximum energy recovery, and the corresponding grid display is shown below.

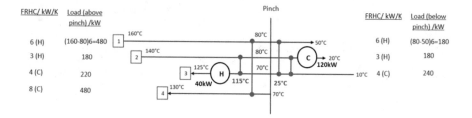

Problem 7

Heat integration is required for the following process streams.

Stream	Flow rate heat capacity [kW/K]	Supply temperature [°C]	Target temperature [°C]
1	18	175	35
2	11	135	85
3	9	45	185
4	11	75	165

Using a minimum approach temperature $\Delta T_{min} = 10$ K,

(a) **Determine the pinch temperatures for the hot and cold streams, as well as the minimum utility duties required.**
(b) **Above the pinch, stream 1 needs to be split. Give reasons why this is the case. Determine the stream matches that would achieve maximum energy recovery and show your results in a grid display.**

Solution 7

(a)

We can determine the pinch temperature and minimum utilities required using an energy cascade diagram. So, we start off by following the standard steps in constructing this diagram.

- Determine shifted temperatures using $\Delta T_{min} = 10$ K
- Label the streams "hot" or "cold"

Stream		W [kW/K]	T_{supply} [$\overset{\circ}{}$C]	T_{target} [$\overset{\circ}{}$C]	$T_{supply}{}^*$ [$\overset{\circ}{}$C]	$T_{target}{}^*$ [$\overset{\circ}{}$C]
1	Hot	18	175	35	$175 - 5 = 170$	$35 - 5 = 30$
2	Hot	11	135	85	130	80
3	Cold	9	45	185	$45 + 5 = 50$	$185 + 5 = 190$
4	Cold	11	75	165	80	170

Next, we group our streams into shifted temperature intervals and determine respective heat capacities and enthalpies for each interval.

T^* intervals	Streams present	$\sum W_C - W_H$	$\Delta H = \Delta T$ $(\sum W_C - W_H)$	$\sum \Delta H$	$\sum - \Delta H$
190–170	3	9	$20(9) = 180$	180	-180
170–130	3, 4, 1	$9 + 11 - 18 = 2$	$40(2) = 80$	$180 + 80 = 260$	-260 (most negative)
130–80	3, 4, 2, 1	$9 + 11 - 11 - 18 = -9$	50 $(-9) = -450$	$180 + 80 - 450 = -190$	190
80–50	3, 1	$9 - 18 = -9$	30 $(-9) = -270$	$180 + 80 -$ $450 - 270 = -460$	460
50–30	1	-18	20 $(-18) = -360$	$180 + 80 - 450 -$ $270 - 360 = -820$	820

The ΔH values refer to the enthalpy changes of the process streams within a specific shifted temperature interval. Therefore, the negative value of this, i.e., $-\Delta H$ would be the enthalpy change of the surroundings external to the process streams. The last column of $\sum -\Delta H$ is used to find the pinch point, which occurs at the lower temperature bound of the interval where the most negative value of $\sum -\Delta H$ is found, i.e., the pinch temperature is 130 °C (shifted) where the most negative value of $\sum -\Delta H = -260$ kW is. The minimum heating utility $Q_{H,\,min}$ is then equivalent to the magnitude of this most negative value. Hence $Q_{H,min} = \mathbf{260}$ **kW**.

The energy cascade diagram showing enthalpy values at each shifted temperature interval is shown below.

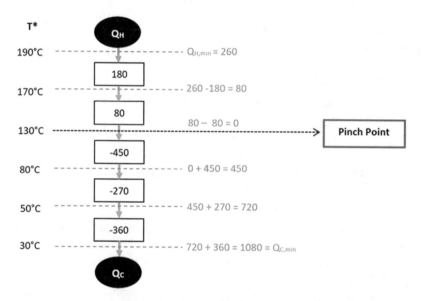

As shown in the diagram, the minimum cooling utility $Q_{C,\,min}$ can be found by going down the cascade starting from the known value of $Q_{H,\,min} = 260$ kW at the top, and working through the cumulative enthalpy changes down the cascade, until we reach the bottom where we compute the final residual enthalpy that exits the last temperature interval, giving us $Q_{C,min} = \mathbf{1080}$ **kW**.

In our energy cascade diagram, we identified our pinch at $T^* = 130\,°C$ (shifted), which means that the actual pinch temperatures are $T_{pinch,hot} = \mathbf{135}\,°\mathbf{C}$ and $T_{pinch,\,cold} = \mathbf{125}\,°\mathbf{C}$.

(b)

Let us construct a simple grid diagram to have an idea of where our pinch point lies in relation to the supply and target temperatures of the four streams. In the diagram below, the flow rate heat capacities (FRHC) of the streams are also indicated.

Pinch (125°C for cold streams, 135°C for hot streams)

Recall that there are a few rules that the sections above, below, and at the pinch should obey respectively.

Below the Pinch (Cold End)

- There is a net heat excess; hence this section is a net heat source. Only cold utilities Q_C can be placed.
- The number of hot streams cannot be less than the number of cold streams, i.e., $N_H \geq N_C$.

  - This is fulfilled as seen in the stream display above.

Above the Pinch (Hot End)

- There is a net heat deficit; hence this section is a net heat sink. Only hot utilities Q_H can be placed.
- The number of cold streams cannot be less than the number of hot streams, i.e., $N_C \geq N_H$.

  - This is also fulfilled as seen in the stream display.

At the Pinch

- On the cold side, flow rate heat capacities of hot and cold streams matched must fulfill $W_H \geq W_C$.

  - This is fulfilled as seen in the stream display ($W_{H,\,1} > W_{C,\,3}$ and $W_{H,\,2} > W_{C,\,3}$).

- On the hot side, flow rate heat capacities of hot and cold streams matched must fulfill $W_C \geq W_H$.

  - Hot stream 1 has a very high heat capacity of 18 kW/K, and would not be able to satisfy this condition $W_C \geq W_H$ since all cold streams above the pinch (streams 3 and 4) have heat capacities less than 18 kW/K.

$$W_{H,\,1} > W_{C,\,3} \quad \text{and} \quad W_{H,\,1} > W_{C,\,4}$$
$$18 > 9 \quad \quad \text{and} \quad \quad 18 > 11$$

- **We need to split stream 1** since this also allows us to split its original heat capacity value into two smaller values, where the sum of both would return the value of 18 kW/K. This helps us fulfill the requirement $W_C \geq W_H$ above the pinch.
- If we split stream 1 into 1a and 1b, whereby $W_{H, 1a} = x$, $W_{H, 1b} = 18 - x$, then since the condition is $W_C \geq W_H$, the constraints on the value of x are as follows:

$$x \leq W_{C, 3} \quad \text{and} \quad 18 - x \leq W_{C, 4}$$
$$x \leq 9 \quad \quad \text{and} \quad 18 - x \leq 11$$
$$\textbf{Therefore, } \mathbf{7 \leq x \leq 9}$$

- Note that by splitting stream 1 above the pinch, we concurrently increase the number of hot streams N_H above the pinch by 1. This is acceptable since we still satisfy the required condition $N_C \geq N_H$ above the pinch after the stream splitting. With a new total of two hot streams and two cold streams, the equality sign of the condition $N_C \geq N_H$ is satisfied.

How do we determine the value of x?
Some general tips in finding an appropriate value of x are as follows:

1. Try selecting a value within the allowable range (in this case, $7 \leq x \leq 9$) that helps us satisfy the full load (and if possible the largest load too) of as many streams as possible in the match.
2. As a rule of thumb, if the allowable range for heat capacity was $7 \leq x \leq 9$, then it typically works well if we pick the lower limit for the split stream represented by x (=7) and the other stream consequently has a heat capacity of $18 - x = 11$.

For better understanding, we will run through a few possible cases for the value of x, to better appreciate how the tips above work.

There are three possible values that x can take and we shall determine which would be the most preferred based on tips 1 and 2 above:

- Case 1: $W_{H, 1a} = x = 7$; $W_{H, 1b} = 11$
- Case 2: $W_{H, 1a} = x = 8$; $W_{H, 1b} = 10$
- Case 3: $W_{H, 1a} = x = 9$; $W_{H, 1b} = 9$

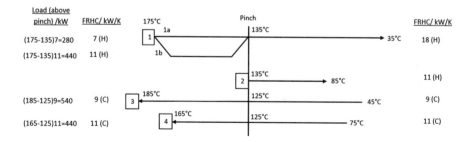

For Case 1, three matches now fulfill the condition of $W_C \geq W_H$ for matches at the pinch on the hot side. They are:

- Hot stream 1a ($W_{H,\ 1a} = 7$) and cold stream 3 ($W_{C,\ 3} = 9$), since $9 \geq 7$
- Hot stream 1a ($W_{H,\ 1a} = 7$) and cold stream 4 ($W_{C,\ 4} = 11$), since $11 \geq 7$
- Hot stream 1b ($W_{H,\ 1b} = 11$) and cold stream 4 ($W_{C,\ 4} = 11$), since $11 \geq 11$ (equality sign fulfilled)

Since there are three possible matches here, we will select the best match which will be the one that is able to satisfy the **full load** of the stream(s) as much as possible (tip 1). The match between streams 1b and 4 best satisfies the full loads of the stream (s) (**both** streams 1b and 4 have **full** loads satisfied) and is therefore preferred over the other two matches.

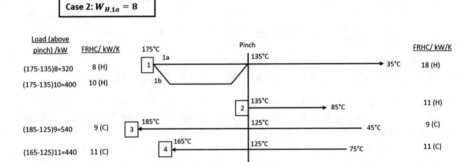

For Case 2, three matches fulfill the condition of $W_C \geq W_H$ for matches at the pinch on the hot side. They are:

- Hot stream 1a ($W_{H,\ 1a} = 8$) and cold stream 3 ($W_{C,\ 3} = 9$), since $9 \geq 8$
- Hot stream 1a ($W_{H,\ 1a} = 8$) and cold stream 4 ($W_{C,\ 4} = 11$), since $11 \geq 8$
- Hot stream 1b ($W_{H,\ 1b} = 10$) and cold stream 4 ($W_{C,\ 4} = 11$), since $11 \geq 10$

Again, there is more than one possible match that fulfills $W_C \geq W_H$, so we will again apply tip 1. We note that none of the matches in Case 2 perform better than the earlier match selected in Case 1 (between streams 1b and 4) as none of the matches in Case 2 can fulfill **both** loads of the streams **fully**.

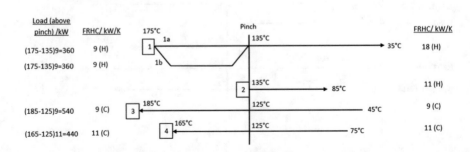

For Case 3, four matches fulfill the condition of $W_C \geq W_H$ for matches at the pinch on the hot side. They are:

- Hot stream 1a ($W_{H,\,1a} = 9$) and cold stream 3 ($W_{C,\,3} = 9$), since $9 \geq 9$ (equality sign fulfilled).
- Hot stream 1b ($W_{H,\,1b} = 9$) and cold stream 3 ($W_{C,\,3} = 9$), since $9 \geq 9$ (equality sign fulfilled).
- Hot stream 1a ($W_{H,\,1a} = 9$) and cold stream 4 ($W_{C,\,4} = 11$), since $11 \geq 9$.
- Hot stream 1b ($W_{H,\,1b} = 9$) and cold stream 4 ($W_{C,\,4} = 11$), since $11 \geq 9$

Like before, since there are multiple matches that fulfill $W_C \geq W_H$, we will apply tip 1 to further select the best match.

We note that similar to Case 2, none of the matches in Case 3 are better than the earlier match in Case 1 (between streams 1b and 4).

Finally, we observe that the best match was found in Case 1, which corresponds to having x take on the lower limit value of the heat capacity range ($=7$) and the other split stream having a heat capacity of $18 - x = 11$ (refer to tip 2).

<u>Confirmed Stream Splitting</u>

We have now confirmed the stream-split and match between hot stream 1b and cold stream 4 at the pinch on the hot side. The original stream 1 with a heat capacity of $W_{H,\,1} = 18$ is now split into two streams, stream 1a with $W_{H,\,1a} = 7$ and stream 1b with $W_{H,\,1b} = 11$. This match at the pinch completely satisfies the full loads of both streams at 440 kW. Since there is a heat transfer of 440 kW from hot stream 1b to cold stream 4, the heat exchanger duty associated with this match is 440 kW.

We have updated our grid display diagram with our first match as shown below.

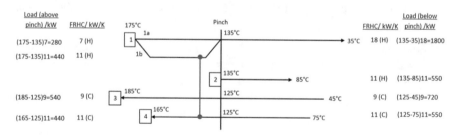

We will now place our next match at the pinch (on the cold side) by finding a match that fulfills $W_H \geq W_C$. The possible matches are as follows. Like before, the best match out of the list is that between streams 2 and 4 since it satisfies the **full** loads of **both** streams.

- Streams 1 and 3
- Streams 1 and 4
- Streams 2 and 3
- Streams 2 and 4—best match that fulfills both streams' loads at 550 kW.

This leaves us with only streams 1 and 3 that are unmatched; hence it is logical to place our second match at the pinch (cold side) between streams 1 and 3. The pairing between 1 and 3 is checked to be allowable since it also obeys the condition $W_H \geq W_C$.

The grid display below is updated with the two new matches at the pinch (cold side).

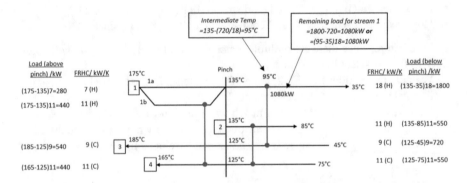

We note that, below the pinch, all streams have their loads completely fulfilled except stream 1, which has a residual load of 1080 kW. In the match between streams 1 and 3, 720 kW of heat was transferred from hot stream 1 to cold stream 3, which completely fulfills stream 3 but leaves stream 1 with a residual load of $1800 - 720 = 1080$ kW.

We found earlier that the minimum cooling utility requirement $Q_{C,\,min} = 1080$ kW; therefore the remaining load can be completely satisfied by placing a cold utility on stream 1. This completes our HEN design for the section below the pinch as shown.

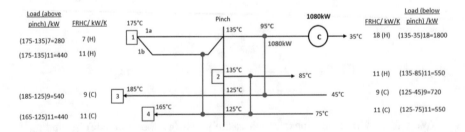

We are now left with the section above the pinch, where we left off with our first match at the pinch between streams 1b and 4 that completely satisfied both streams' loads.

It makes sense to match the other two remaining streams 1a and 3, which we check to fulfill $W_C \geq W_H$ at the pinch (hot side). With this match, stream 1a's 280 kW load will be completely fulfilled while stream 3 will be left with a remaining load of $540 - 280 = 260$ kW.

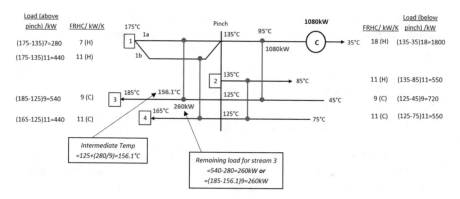

All streams above the pinch are now completely fulfilled, except for stream 3 with a residual load of 260 kW. We found earlier that the minimum heating utility requirement $Q_{H, \, min} = 260$ kW; therefore the remaining load can be completely satisfied by placing a hot utility on stream 3.

Combining our results above and below the pinch, we have our complete HEN design for maximum energy recovery in the grid display below.

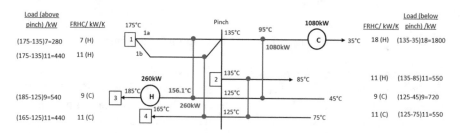

Problem 8

You are provided with the following grid display showing process streams above the pinch. The hot pinch temperature, supply temperature of hot stream 1, as well as target temperatures of cold streams 2 and 3 are indicated.

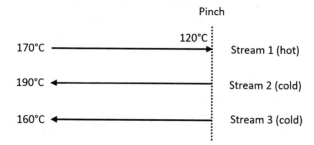

Given that the flow rate heat capacities of streams 1, 2, and 3 are 0.7, 0.4, and 0.6 kW/K, respectively, and the minimum approach temperature $\Delta T_{min} = 10$ K,

complete the grid display by filling in missing details. Then show stream matches that are possible in order to achieve heat integration.

Solution 8

Given the hot pinch temperature, we can find the corresponding cold pinch temperature using the given value of $\Delta T_{min} = 10$ K.

$$T_{pinch,cold} = T_{pinch,hot} - \Delta T_{min} = 120 - 10 = 110°C$$

The cold pinch temperature applies for all cold streams, i.e., streams 2 and 3. We can complete the grid display by filling in the cold pinch temperature and heat capacities for the streams as shown below.

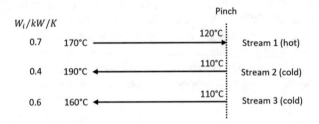

We can now begin placing stream matches, starting at the pinch.

Conditions at the Pinch (Hot Side)

The condition required for matches at the pinch on the hot side is $W_C \geq W_H$. However, both cold streams 2 and 3 have heat capacities that are smaller than that of hot stream 1. Therefore, we will need to split stream 1 such that its heat capacity ($W_{H,1} = 0.7$) is also divided into two smaller values in the two split streams (e.g., 1a and 1b). Each of the split streams must then be able to match with streams 2 and 3 and obey the rule $W_C \geq W_H$.

One way to split stream 1 is such that $W_{H,1a} = 0.5$ and $W_{H,1b} = 0.2$. We are then able to match streams 1a with 3 and 1b with 2. Note that we cannot match streams 1a and 2 as that would again violate the requirement $W_C \geq W_H$.

[As good practice, we should also check that we have fulfilled the other condition on number of streams for the section above the pinch, $N_C \geq N_H$. In this case, we have two hot streams (1a and 1b) and two cold streams, so the equality sign of the condition is fulfilled.]

Heat Exchanger (HXer) Duty and Residual Loads

In the match between streams 1a and 3, hot stream 1a transfers 25 kW of heat to cold stream 3; therefore the HXer duty representing this match is also 25 kW. This match leaves stream 1's load above the pinch completely fulfilled, while stream 3 is left with a 5 kW ($=30 - 25$) heating load remaining. Therefore, a hot utility of 5 kW may be placed on stream 3 to fulfill this remaining load.

In the match between streams 1b and 2, hot stream 1b transfers 10 kW of heat to cold stream 2; therefore the HXer duty representing this match is 10 kW. This match leaves stream 1b's load completely fulfilled while stream 2 has a remaining heating load of 22 kW (=32 − 10) that can be met with a hot utility of 22 kW placed on stream 2.

The matches and utility placements are shown in the diagram below.

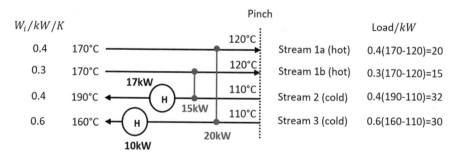

Other Possible HEN Designs

In the worked solution above, we have defined the split streams as 1a and 1b, with heat capacity values of 0.5 and 0.2 kW/K respectively, such that they fulfill the required condition at the pinch. We then obtained the HEN design whereby two HXers (10 and 25 kW) and two hot utilities (5 and 22 kW) were used.

One may note that there is more than one way to split stream 1 and still fulfill the condition at the pinch. Let us examine one such alternative case below.

Alternative stream-split: $W_{H,1a} = 0.4$ and $W_{H,1b} = 0.3$

With this split, we can still fulfill the required condition $W_C \geq W_H$ at the pinch (hot side) with the following stream matches and hot utility configurations:

Configuration 1:

The grid display for Configuration 1 is shown below:

$W_i/kW/K$						Load/kW
0.4	170°C			120°C	Stream 1a (hot)	0.4(170-120)=20
0.3	170°C	17kW		120°C	Stream 1b (hot)	0.3(170-120)=15
0.4	190°C	H	15kW	110°C	Stream 2 (cold)	0.4(190-110)=32
0.6	160°C	H	20kW	110°C	Stream 3 (cold)	0.6(160-110)=30
		10kW				

- Stream 1a and 3 ($W_{C,\,3} \geq W_{H,\,1a}$) matched using a 20 kW HXer.
- Stream 1b and 2 ($W_{C,\,2} \geq W_{H,\,1b}$) matched using a 15 kW HXer.

- One hot utility of 17 kW duty placed on stream 2.
- One hot utility of 10 kW duty placed on stream 3.

Configuration 2:
The grid display for Configuration 2 is shown below:

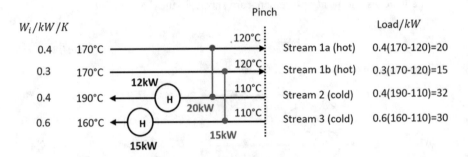

- Stream 1a and 2 ($W_{C, 2} \geq W_{H, 1a}$, equality sign fulfilled) matched using a 20 kW HXer.
- Stream 1b and 3 ($W_{C, 3} \geq W_{H, 1b}$) matched using a 15 kW HXer.
- One hot utility of 12 kW duty placed on stream 2.
- One hot utility of 15 kW duty placed on stream 3.

An interesting observation to note here is that regardless of how we split stream 1, the total enthalpy load fulfilled by stream matches is 35 kW while the total enthalpy load fulfilled by external hot utilities is 27 kW. In other words, the extent of heat integration and energy recovery is not affected by the specific ratio of heat capacities from the split. This makes sense since our system has not changed with the split, as the total heat capacity of the two split streams still sums to that of the original stream 1, with no other changes to system parameters (i.e., supply and target temperatures of streams and pinch temperatures remain unchanged).

Chapter 3
Euler's Theorem and Grand Composite Curves

Problem 9

Euler's network theorem may be applied to heat exchanger design whereby the number of heat exchanger units U may be related to the total number of streams N, the number of loops in the system L, and the number of subsystems S as shown below.

$$U = N + L - S$$

(a) **Design a heat exchanger network for the following process streams for maximum energy recovery (MER) and hence determine the value of $U_{min, MER}$. Show your HEN design on a grid diagram, indicating clearly all process streams above and below the pinch and any relevant intermediate temperatures at the entrances and exits of the heat exchanger units. You may assume $\Delta T_{min} = 10$ K.**

Stream label	Heat capacity [kW/K]	Supply temperature [°C]	Target temperature [°C]
1	11	200	40
2	3	170	80
3	9	30	160
4	19	90	140

(b) **Using Euler's network theorem, determine $U_{min, overall}$, and compute the maximum reduction in the number of heat exchanger units that is possible from the HEN designed in part a.**

X. W. Ng, *Concise Guide to Heat Exchanger Network Design*, https://doi.org/10.1007/978-3-030-53498-1_3

(c) **You are asked to reduce the number of heat exchangers by 1 from your HEN design in part a. Find out new utility duties required for this case.**

(d) **Draw the grand composite curve and find the minimum hot utility temperature and maximum cold utility temperature.**

Solution 9

(a)

We first label the streams either "hot" or "cold" by observing their supply and target temperatures. We then compute shifted temperatures T^* for the given supply and target temperatures by applying $\Delta T_{min} = 10$ K.

Stream label	W [kW/K]	T_{supply}^* [°C]	T_{target}^* [°C]
1 (hot)	11	$200 - 5 = 195$	$40 - 5 = 35$
2 (hot)	3	$170 - 5 = 165$	$80 - 5 = 75$
3 (cold)	9	$30 + 5 = 35$	$160 + 5 = 165$
4 (cold)	19	$90 + 5 = 95$	$140 + 5 = 145$

Next, we group the streams into shifted temperature intervals and compute for each interval the total heat capacity $\sum W_C - W_H$, enthalpy change ΔH, and cumulative enthalpy change $\sum \Delta H$ as shown below.

T^* intervals	Streams present	$\sum W_C - W_H$ [kW/K]	ΔH [kW] $= (\Delta T)$ $(\sum W_C - W_H)$	$\sum \Delta H$ [kW]	$-\sum \Delta H$ [kW]
195–165	1(H)	-11	$30(-11) = -330$	-330	330
165–145	1(H), 2(H), 3(C)	$-11 - 3 + 9 = -5$	$20(-5) = -100$	$-330 - 100 = -430$	430
145–95	1(H), 2(H), 3(C), 4(C)	$-11 - 3 + 9 + 19 = 14$	$50(14) = 700$	$-330 - 100 + 700 = 270$	-270 (most negative)
95–75	1(H), 2(H), 3(C)	$-11 - 3 + 9 = -5$	$20(-5) = -100$	$-330 - 100 + 700 - 100 = 170$	-170
75–35	1(H), 3(C)	$-11 + 9 = -2$	$40(-2) = -80$	$-330 - 100 + 700 - 100 - 80 = 90$	-90

As explained in earlier problems (refer to the chapter on energy cascade and pinch analysis for more details), the magnitude of the most negative value in the last column of $-\sum \Delta H$ will be equivalent to our minimum heating utility, i.e., $Q_{H, min} = 270$ kW. Also, the pinch point can be immediately identified to be the lower limit of the interval containing the most negative value of $-\sum \Delta H$, i.e., $T_{pinch} = 95°C$ (shifted) or $T_{pinch, hot} = 100°C$ and $T_{pinch, cold} = 90°C$. We may then construct our energy cascade to determine the minimum cooling utility, $Q_{C, min}$. For maximum energy recovery, utility requirements are minimum as $\Delta T_{min} = 10$ K is met at the pinch.

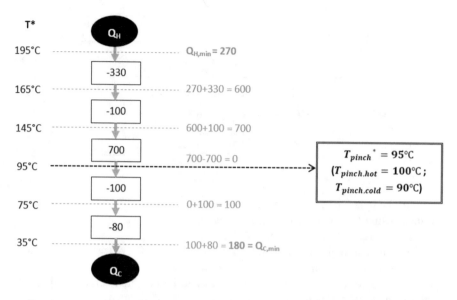

From our energy cascade, we can find our minimum cooling utility $Q_{C,min}$ = 180 kW by following through with the cascade from top to bottom, summing the enthalpies each time we pass through a temperature interval, as shown in the diagram above.

In summary, we have found $T_{pinch,\,hot}$ = $100\,°C$, $T_{pinch,\,cold}$ = $90\,°C$, $Q_{H,min}$ = 270 kW, and $Q_{C,\,min}$ = 180 kW. We may now use this information to construct our grid display as shown above, and design our heat exchanger network (HEN) for MER.

Matches at the Pinch

At the pinch on the cold side, flow rate heat capacities of matched streams must fulfill $W_H \geq W_C$. Only streams 1 and 3 fulfill this condition, so we can directly indicate this match in our grid display and calculate any residual loads and intermediate temperatures.

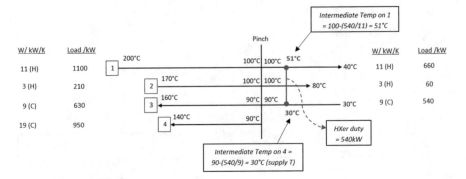

From the match between 1 and 3, stream 3's load of 540 kW will be completely fulfilled, while stream 1 would have a remaining load of 120 kW (=660 − 540). Separately, stream 2 still retains its full load of 60 kW.

Therefore, the total residual load below the pinch is 180 kW (=120 + 60). Since this value is exactly equivalent to the minimum cooling utility $Q_{C, \text{min}} = 180$ kW found earlier, we can fulfill the rest of stream 1 and 2's cooling loads by adding cold utilities, i.e., a 120 kW cold utility to stream 1 and a 60 kW cold utility to stream 2 as shown below.

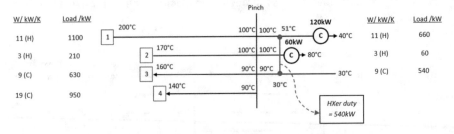

We have completed our HEN design for the section below the pinch and may now work on matching streams at the pinch on the hot side. This time, we require flow rate heat capacities of matched streams to fulfill $W_C \geq W_H$. This means that the following matches are all possible:

- Hot stream 2 and cold stream 3 ($W_{C, 3} \geq W_{H, 2} \rightarrow 9 \geq 3$)
- Hot stream 2 and cold stream 4 ($W_{C, 4} \geq W_{H, 2} \rightarrow 19 \geq 3$)
- Hot stream 1 and cold stream 4 ($W_{C, 4} \geq W_{H, 1} \rightarrow 19 \geq 11$)

Note that in order to obey the condition ($W_C \geq W_H$), stream 1 can only match with 4 but not 3. Therefore, we should reserve stream 4 to match with stream 1, and this indirectly helps us confirm our next match between streams 2 and 3 since they are the remaining unmatched hot and cold streams. Moreover, we confirm that a match between two and three still fulfills the required $W_C \geq W_H$ at the pinch.

We may now update our grid display with the abovementioned matches at the pinch (hot side), starting with the match between streams 1 and 4 as shown. This match uses a heat exchanger (HXer) of 950 kW duty, since 950 kW of heat is transferred from hot stream 1 to cold stream 4 in the HXer.

This match also leaves stream 1 with a residual load of $1100 - 950 = 150$ kW. The intermediate temperature of stream 1 entering the HXer is calculated to be $186\,°C$, while the temperature of stream 4 exiting the HXer is calculated to be $140\,°C$ (equivalent to its target temperature since its load is completely fulfilled by this match).

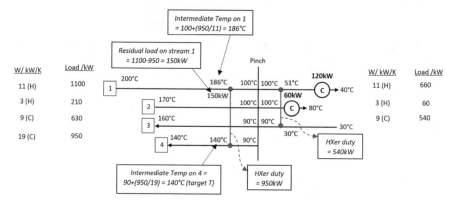

Next, we add to our grid display the match between streams 2 and 3, also at the pinch (hot side). This match gives us a HXer duty of 210 kW, since 210 kW of heat is transferred from hot stream 2 to cold stream 3 in the HXer.

This match also leaves stream 3 with a residual load of $630 - 210 = 420$ kW. The intermediate temperature of stream 3 leaving the HXer is calculated to be $113\,°C$, while the temperature of stream 2 entering the HXer is calculated to be $100\,°C$ (equivalent to the hot pinch temperature since its load is completely fulfilled by this match).

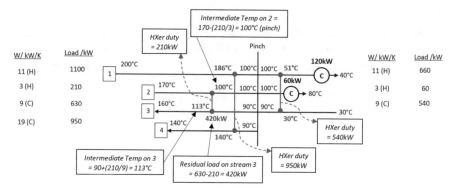

We may do a "stock-take" at this point, by checking our current total residual load. We still have a residual load of 150 kW (i.e., this amount of cooling still required) on stream 1 and 420 kW (i.e., this amount of heating still required) on stream 3. We recall that one of the rules for the section above the pinch is that we can only place hot utilities and not cold utilities. Therefore, we will need to place more stream matches in order to fulfill the residual cooling load of stream 1.

Matches Away from the Pinch

The next logical match is between streams 1 and 3 since they are the only streams left with residual loads. Note that at this point, we have moved away from the pinch, and it is no longer necessary to obey the $W_C \geq W_H$ condition for streams 1 and 3 in this match.

- Stream 1—residual load of 150 kW (cooling required)
- Stream 3—residual load of 420 kW (heating required)

This match gives us a HXer duty of 150 kW, since 150 kW of heat is transferred from hot stream 1 to cold stream 3 in the HXer. This match also leaves stream 3 with a residual load of $420 - 150 = 270$ kW. The intermediate temperature of stream 3 leaving the HXer can be calculated to be 130 °C, using the earlier-found intermediate temperature 113 °C as the inlet temperature to this HXer. The temperature of stream 1 entering this HXer can be calculated to be 200 °C (equivalent to its supply temperature since its residual load is completely fulfilled by this match).

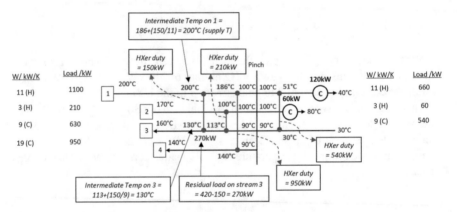

We now observe that all streams have completely fulfilled their loads except stream 3, which still has a residual load of 270 kW. This value is exactly equivalent to the minimum heating utility $Q_{H, min} = 270$ kW found earlier; therefore we can now place a hot utility on stream 3, which will fulfill the rest of the residual load above the pinch. The final grid display combining our results both above and below the pinch is shown below.

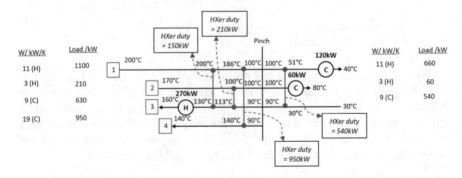

Our HEN design for MER consists of four HXers in total, three above the pinch and one below the pinch. We also have one hot utility above the pinch and two cold utilities below the pinch. The results are summarized below with the respective duties.

Heat exchanger duties / kW	Streams involved	Utility duties	Stream placed
Above pinch		**Above pinch (heating)**	
950	1, 4	270kW	3
210	2, 3	**Below pinch (cooling)**	
150	1, 3	120	1
Below pinch		60	2
540	1, 3		

Euler's network theorem is often used in analyzing HEN designs and is shown by the equation below.

$$U = N + L - S$$

- **U refers to the number of HXer units.** This includes the number of stream matches and external utilities.

  - A stream match essentially represents heat transfer from a hot process stream to a cold process stream. This is achieved via a HXer that brings both streams close to each other for heat transfer over the heat exchange area.
  - Utilities are also heat exchange systems. For example, a hot utility could be high-temperature steam, which transfers its heat to a process stream that requires cooling. A cold utility could be a coolant fluid such as cooling water at 20 °C or lower, which is able to receive heat from a hot process stream.
  - Note that if we have two cold (or hot) utilities placed on **separate** streams, we will count them separately, i.e., a contribution of 2 to the value of U.

- **N refers to the number of streams and utilities.**

  - It is important to note that the way we count the number of utilities for N is different from that for U.
  - If we have multiple hot (or cold) utilities placed on **separate** streams, we will count them **once** as long as they are the **same type** of utility (i.e., hot or cold).
  - For example, if we had one hot utility on stream A, one hot utility on stream B, one cold utility on stream C, and one cold utility on stream D, then the value of N will be 6 and not 8.

Correct computation: $N = 6 = 4$ (streams) $+ 2$ (one hot utility and one cold utility).

Incorrect computation: $N = 8 = 4$ (streams) $+ 4$ (two hot utilities and two cold utilities).

- **L refers to the number of loops or closed paths in the system**, whereby a closed path is formed by connecting stream matches and/or utilities in a complete loop.

  - It is optimal to achieve $L = 0$ to avoid having excessive HXer units.
  - When we "break" a loop, we save a HXer unit.

- **S refers to the number of individual subsystems**. A subsystem consists of a group of stream matches that can form separate HENs via enthalpy balance. In most cases, S is assumed to be equivalent to 1.

From our earlier HEN design for MER, we note that:

Above the Pinch

- There are three stream matches and one hot utility; therefore $U = 4$.
- There are four process streams (i.e., streams 1, 2, 3, and 4) and one hot utility; therefore $N = 4 + 1 = 5$.
- There are no loops observed, $L = 0$. When there are no loops, the number of HXers is also minimized, i.e., $U = 4$ is the minimum for the section above the pinch.
- $S = 1$ for considering just the section above the pinch.
- $U = N + L - S \rightarrow 4 = 5 + 0 - 1$.

Below the Pinch

- There is one stream match and two cold utilities; therefore $U = 3$.
- There are three process streams (i.e., streams 1, 2, and 3) and one cold utility; therefore $N = 3 + 1 = 4$.
- There are no loops observed, $L = 0$. When there are no loops, the number of HXers is also minimized, i.e., $U = 3$ is the minimum for the section below the pinch.
- $S = 1$ for considering just the section below the pinch.
- $U = N + L - S \rightarrow 3 = 4 + 0 - 1$.

In Total

$$U_{\min,\text{MER}} = U_{\min,\text{MER,above pinch}} + U_{\min,\text{MER,below pinch}} = 4 + 3 = 7$$

Considering the overall system, we have a total number of 7 HXer units for MER; this corresponds to having 4 HXer units, 1 hot utility, and 2 cold utilities.

(b)

We now need to determine the value of $U_{\min,\text{ overall}}$ but before that, let us first discuss the difference between $U_{\min,\text{ overall}}$ and $U_{\min,\text{ MER}}$.

Difference Between $U_{min,overall}$ and $U_{min,MER}$

When we designed our HEN for MER in part (a), we placed matches above and below the pinch separately using pinch analysis since ΔT_{min} had to be fulfilled at the pinch. In doing so, we were in fact optimizing (i.e., minimizing) the number of HXers separately in these two sections. However, we do not necessarily optimize the number of HXers for the overall system. It is possible that $U_{min, \, overall} \neq U_{min, \, MER}$.

In fact in this problem, we will have $U_{min, \, overall} < U_{min, \, MER}$. This is because when we placed matches separately for the two sections (above and below the pinch), we introduced loops **across** the pinch, so L is no longer zero for the overall system, $L > 0$ instead. The optimum (or minimum) number of HXer units for the entire system $U_{min, \, overall}$, requires $L = 0$ overall. On the other hand, $U_{min, \, MER}$ meant that $L = 0$ separately for the sections above and below the pinch, but $L > 0$ overall.

The difference between the two values, i.e., $U_{min, \, MER} - U_{min, \, overall}$, therefore represents the <u>scope for reducing the number of HXer units used.</u>

Scope for Reducing the Number of HXers Used

<u>Method 1</u>

When we have loops in a system ($L > 0$), it means we have scope to reduce the number of HXers used. When we break loop(s), we are effectively removing HXer(s), that is why when $L = 0$, we have the optimal (minimum) number of HXers.

Using the value of $U_{min, \, MER} = 7$ found in part a, we can **back-deduce the number of loops** we have in the overall system by applying Euler's network equation on the system as a whole (instead of splitting into sections above and below the pinch like before).

Considering the overall system as one, we have

- Four process streams, one hot utility and two cold utilities (counted as one cold utility in N as mentioned earlier). Therefore $N = 4 + 1 + 1 = 6$.
- $S = 1$ for this overall system considered in entirety.

$$U_{min,MER} = N + L - S$$
$$7 = 6 + L - 1$$
$$7 = 5 + L$$
$$L = 2$$

- We have found that our **MER design introduced two loops** into the overall system. This indirectly tells us that we have scope to remove two HXers from $U_{min, \, MER} = 7$ to $U_{min, \, overall} = 5$.

<u>Method 2:</u>

We can also determine the scope for reducing the number of HXers by directly computing $U_{min, \, overall}$ and comparing it with the value of $U_{min, \, MER} = 7$ earlier found in part (a).

Again, considering the overall system as one, we have

- We have $N = 6$ (same as in Method 1).
- We will have no loops if we are finding the minimum number of HXer units; therefore $L = 0$.
- We have $S = 1$ for the overall system (same as in Method 1).
- $U_{min, \, overall} = N + L - S = 6 + 0 - 1 = 5$.

The scope for reducing the number of HXers is again $U_{min, \, MER} - U_{min, \, overall} = 7 - 5 = 2$.

"Compromise" in Reducing U

As we found using both Methods 1 and 2, the minimum number of HXers needed for the overall system, such that we will still be able to achieve required target temperatures for all streams, is 5.

In part a, we found that we require seven HXers for maximum energy recovery (MER).

In reducing the number of HXers from 7 to 5, our system will no longer be at MER. This means that we will have a reduced extent of heat integration between the streams (or smaller amount of energy recovery). The compromise in this is that we will then need to increase our external utility duties, in order to still ensure all streams achieve their target temperatures and fulfill their required heating and cooling loads.

(c)

In removing HXer units, we can use a simple rule of thumb, which is to first remove the one with the smallest duty. The steps are summarized as follows:

1. **Identify a loop** within your HEN design for MER that:

   - Crosses the pinch (i.e., extends from the section above the pinch, to the section below the pinch).

     - Since no loops are present in the separate sections above the pinch and below the pinch (see part (a)), any loops present in the overall system must exist across the pinch point.

   - Does not pass through utilities (i.e., formed only by stream matches).

     - This ensures that we can "break" loops (i.e., reduce the number of loops) by removing HXers for stream matches.

2. **Remove the HXer unit with the smallest duty** within the identified loop **and add back the duty of this removed HXer to the remaining HXer in the loop**.

   - In general, a loop is defined as any closed path formed by connecting stream matches and/or utilities in a complete cycle. Our loop here does not pass through utilities (see step 1); therefore the "removed duty" is added back to the remaining HXer in the loop (as there are no utilities).

- Note that since we are removing HXers now, the system is no longer at MER, and the earlier pinch temperatures are no longer valid.

3. **Recalculate new intermediate stream temperatures** after removal of the HXer in step 2.

4. **Identify violations** that arise as a result of steps 1–3, specifically the following:

   - Check that ΔT_{min} is maintained throughout the system.
   - Check that the **second law of thermodynamics** is obeyed, i.e., **heat flows from hot to cold** and not vice versa.

5. **Correct any violations identified in step 4 by increasing utility duties** and determine the new utility duties required for the case of $U_{min,\,overall}$.

Let us now apply the above steps to our HEN design obtained in part (a).

Steps 1 and 2:

We note that a loop that crosses the pinch but does not pass through utilities is the one marked in yellow below. In this loop, the HXer with the smallest duty is the one above the pinch, between streams 1 and 3 with a duty of 150 kW. We now remove this unit and add back its duty to the remaining HXer within this loop, i.e., the HXer of 540 kW duty located below the pinch, between streams 1 and 3. The new duty for this HXer therefore becomes 690 kW (=540 + 150).

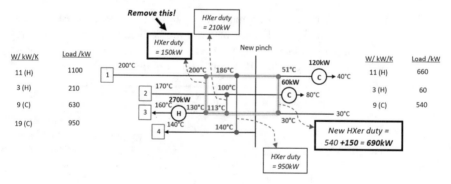

Step 3:

We now proceed to recalculate new intermediate temperatures as a result of the removal of a HXer. The new intermediate temperatures are shown in blue in the diagram below.

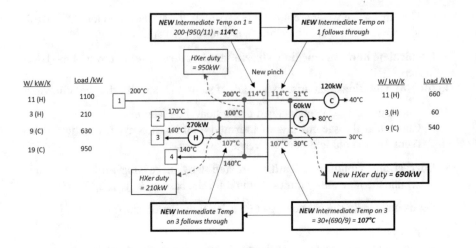

Step 4:

This next step requires us to check for violations.

(a) Minimum approach temperature, ΔT_{min}, should be maintained throughout the system.

- There is a violation of this condition near to the HXer between streams 1 and 3 below the pinch. The difference between the hot and cold stream temperatures at the pinch should be maintained at 10 K (i.e., ΔT_{min}). However, it is now $7°C$ or 7 K which is less than 10 K and hence not possible.

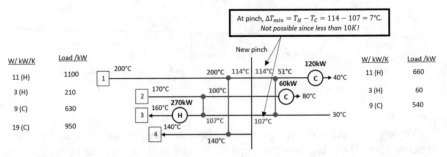

(b) The second law of thermodynamics requires that heat flows from hot to cold and not vice versa.

- There is a violation of this condition for the HXer between streams 2 and 3 above the pinch since the hot stream temperature is lower than the cold stream temperature. This cannot be the case for the correct heat transfer direction, indicated Q below from a hot stream to a cold stream.

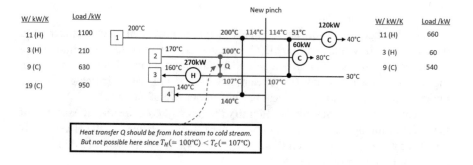

Heat transfer Q should be from hot stream to cold stream.
But not possible here since $T_H(= 100°C) < T_C(= 107°C)$

Step 5:

We may now correct the violations identified in step 4, starting with the violations in ΔT_{min}.

Correcting ΔT_{min} at the Pinch (Cold Side)

To fix the ΔT_{min} violation for the match between streams 1 and 3 at the pinch (cold side), we enforce $\Delta T_{min} = 10$ K $= T_{H, pinch} - T_{C, pinch}$, so the temperature of cold stream 3 at the pinch should be $T_{C, pinch} = 114 - 10 = 104°C$ instead of the earlier $107°C$. This new value for $T_{C, pinch}$ is now updated for cold stream 3 in the diagram below, noting that the same $T_{C, pinch}$ carries forward to **both** the hot and cold sides of the pinch as shown in red.

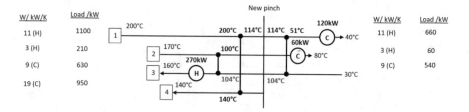

Correcting ΔT_{min} at the Pinch (Hot Side)

Similarly, for the match between streams 2 and 3, we can determine the correct hot pinch temperature on stream 2 by enforcing $\Delta T_{min} = 10$ K. This gives us a new hot pinch temperature $T_{H, pinch} = 104 + 10 = 114°C$. This value is updated for hot stream 2 in red in the diagram below.

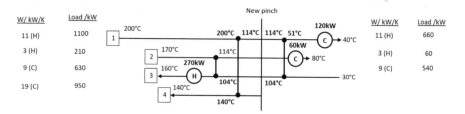

Correcting Utility Duties

There is often a trade-off between heat integration between process streams and the external utility duties required. The greater the extent of heat integration, the smaller the external utility duties required. Therefore, it is logical to expect that with the removal of a HXer unit (for stream match), we will see an increase in utility duties required, so as to restore heat balance in our overall system.

To correct utility duties, we first **identify the hot and cold utilities at each end** of the stream running through the "faulty" HXer ("faulty" defined as the HXer that is affected by the ΔT_{min} violation described in step 4). We then mark out this continuous heat flow path from end-to-end.

In this problem, we have two "faulty" HXers earlier identified in step 4, they are:

- 690 kW HXer below the pinch, between streams 1 and 3—heat path is marked in yellow below.
- 210 kW HXer above the pinch, between streams 2 and 3—heat path is marked in green below.

"Faulty" HXer Between Streams 1 and 3

- **To correct this HXer and its associated utility duties, we add an unknown value y to the utility duties**. This means that the cold utility on stream 1 becomes $(120 + y)$ kW while the hot utility on stream 3 becomes $(270 + y)$ kW. The addition made to utility duties is related to the abovementioned trade-off between removing HXers and increasing utility duties.

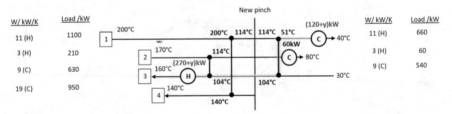

- Since we have added y kW to each of the hot and cold utilities linked to this faulty HXer, we will need to reduce the duty of this HXer by the same amount y kW, in order to maintain energy balance. Recall that the increase in utility duty will have a direct implication on a decrease in heat integration (heat transfer between streams).

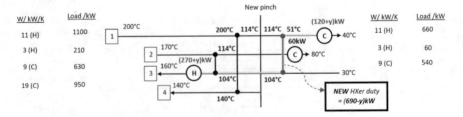

- We are now able to determine the value of y by doing a heat balance on stream 3 near the cold end using its supply temperature of 30 °C and heat capacity denoted W_3.

$$104 - 30 = \frac{690 - y}{W_3}$$

$$74 = \frac{690 - y}{9}$$

$$y = 24$$

Using this value of y, we can now find the new duties for the HXer, hot utility on stream 3, and cold utility on stream 1:

Corrected HXer duty $=690 - y = 690 - 24 = 666$ kW
Hot utility on stream 3 $=270 + y = 270 + 24 = 294$ kW
Cold utility on stream 1 $=120 + y = 120 + 24 = 144$ kW

- We have now corrected the first faulty HXer with the updated grid display as shown below.

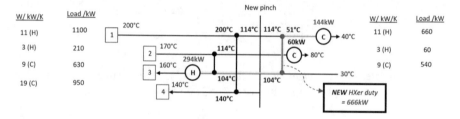

"Faulty" HXer Between Streams 2 and 3

- To correct the second faulty HXer, we mark out the continuous path that it forms with the hot utility on one end (above the pinch) and the cold utility on the other end (below the pinch). This path is marked in green below where the hot utility on stream 3, HXer (between streams 2 and 3), and cold utility on stream 2 are linked in this path.

[As a side note, it will not matter if the path was marked like the one shown above or below. They will give the same results as long as the path passes through the same elements.]

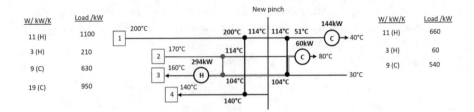

- As before, we check for heat balance along the green path. To do so, we again **add an amount x (defined "x" here to differentiate from the earlier "y") to the utility duties.** This means that the cold utility on stream 2 becomes $(60 + x)$ kW while the hot utility on stream 3 becomes $(294 + x)$ kW. At the same time, we will need to reduce the duty of the "faulty" HXer between streams 2 and 3 by the same amount x kW. This is shown in the diagram below.

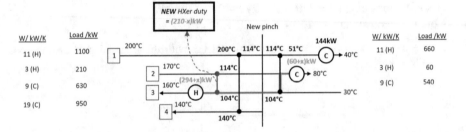

- We can determine the value of x by doing a heat balance on stream 2 near the hot end, using its supply temperature of 170 °C and heat capacity denoted W_2.

$$170 - 114 = \frac{210 - x}{W_2}$$

$$56 = \frac{210 - x}{3}$$

$$x = 42$$

Using this value of x, we can now find the new duties for the HXer, hot utility on stream 3, and cold utility on stream 2:

Corrected HXer duty $= 210 - x = 210 - 42 = 168$ kW
Hot utility on stream 3 $= 294 + x = 294 + 42 = 336$ kW
Cold utility on stream 2 $= 60 + x = 60 + 42 = 102$ kW

- We have now corrected the second faulty HXer as shown below.

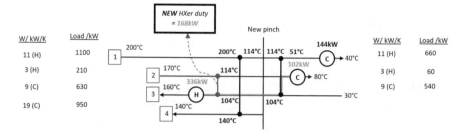

- Combining all our results on the correct pinch temperatures, HXer duties, and utility duties, we obtain our final grid display as shown below.

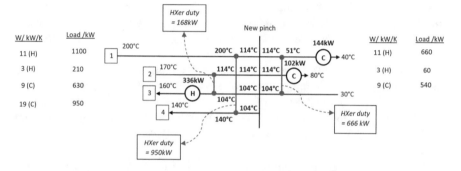

In this HEN design, we observe that when we removed a HXer (for a stream match), it led to increased utility duties (both hot and cold). The hot utility on stream 3 increased from $Q_H = 270$ kW to $Q_H = 336$ kW, while the total cold utility duty (on streams 1 and 2) increased from $Q_C = 180$ kW (where $Q_{C,\,1} = 120$ kW, $Q_{C,\,2} = 60$ kW) to $Q_C = 246$ kW (where $Q_{C,\,1} = 144$ kW, $Q_{C,\,2} = 102$ kW).

Checking That Overall Heat Balance Is Maintained (First Law of Thermodynamics)

One way to double-check that our new HEN design fulfills heat balance is by checking that the **net heat change for our overall system remains constant** regardless of our HEN design.

The supply and target temperatures of process streams do not change; therefore, the heating and cooling loads of our streams do not change, even if we vary our HEN design.

We can calculate the net heat change (could be a net deficit or net gain, depending on the streams defined for the system) of our system by computing the heating and cooling loads of the individual process streams, then summing them up.

Process streams	W [kW/K]	T_{supply} [$^\circ$C]	T_{target} [$^\circ$C]	ΔT [$^\circ$C] $= T_{\text{target}} - T_{\text{supply}}$	ΔH [kW] $= \Delta T \cdot W$
1 (hot)	11	200	40	$40 - 200 = -160$	-160 $(11) = -1760$
2 (hot)	3	170	80	$80 - 170 = -90$	$-90(3) = -270$
3 (cold)	9	30	160	$160 - 30 = 130$	$130(9) = 1170$
4 (cold)	19	90	140	$140 - 90 = 50$	$50(19) = 950$
					$\sum \Delta H = 90$

From the table above, we note that our system is a **net heat sink of 90 kW** (i.e., overall system requires heat input). A net positive value for ΔH of process streams means that their total cooling load outweighs their total heating load; therefore the system will **require a heat input of 90 kW** to achieve energy balance. In other words, if there was no heat input of 90 kW supplied to the system, the streams would not be able to successfully reach their target temperatures as there will be a shortfall of 90 kW heat.

This heat input is in fact provided by the external utilities. We found earlier that the total external heating utility duty $Q_H = 336$ kW and the total external cooling utility duty $Q_C = 246$ kW. Since $Q_H - Q_C = 336 - 246 = 90$ kW, the net utility duty is equivalent to a 90 kW net heat supply to the system. This agrees with our earlier analysis that the system is a net heat sink of 90 kW.

(d)

The Grand Composite Curve (GCC) is a plot of temperature against enthalpy. Unlike the composite curves which are plotted as a pair of hot and cold composite curves (refer to Problem 6, part (b) for details), the GCC is a single plot that considers all streams (hot and cold) within each temperature interval. In addition, the GCC uses **cumulative** enthalpy values (cascaded heat) derived from the energy cascade and therefore takes into consideration enthalpies from external utility duties.

We may use the values found earlier in our energy cascade diagram from part (a) (copied below) to tabulate a series of points for drawing the GCC curve.

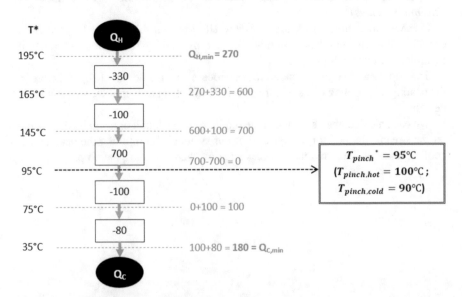

In the table below (left), the first column indicates **shifted temperatures** and the second column indicates total enthalpy cascaded down from a particular temperature. The GCC plot for the data is shown below (right). It is worth noting that the GCC curve has zero enthalpy at the pinch point. This means that there is no heat transfer across the pinch, since the pinch point separates the overall system into two decoupled sections, above and below the pinch respectively.

$T^*/°C$	H/kW
195	270
165	600
145	700
95	0
75	100
35	180

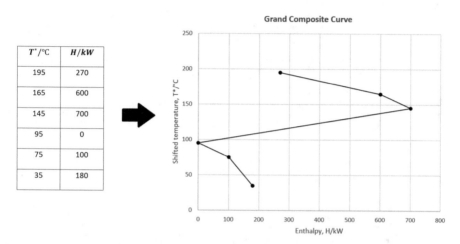

To find the minimum hot utility temperature, we draw a horizontal line from the vertical "Temperature" axis to the data point representing the hot utility. Then, we find out how "low" we can shift this horizontal line downwards while maintaining the same hot utility duty. By doing this, we find that we reach the **minimum possible hot utility temperature of 119.3 °C (or shifted temperature of 114.3 °C)** at the same enthalpy value of 270 kW.

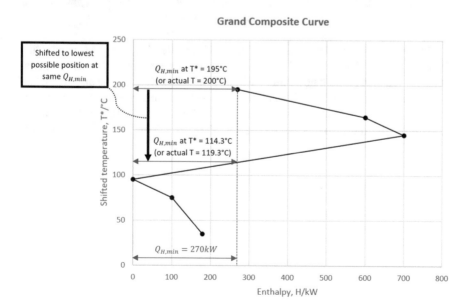

What Happens at Other Temperatures?

We note that $T = 119.3\ °C$ (or $T^* = 114.3\ °C$) is the lowest temperature that the hot utility of 270 kW can be supplied. If we shift the horizontal line to a position where $200 > T > 119.3\ °C$, we would over-supply heat, $Q_H > 270$ kW. If we shift the horizontal line to a position where $T < 119.3\ °C$, we would under-supply heat with $Q_H < 270$ kW. Both do not tally with the intended $Q_{H,\ min} = 270$ kW for maximum energy recovery.

Grand Composite Curve

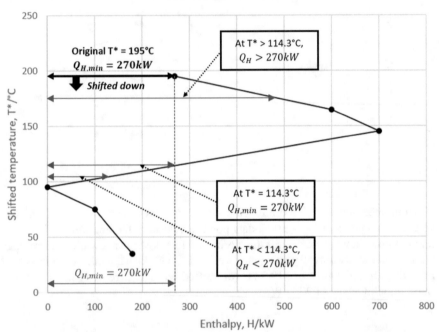

Similarly, to find the maximum cold utility temperature, we draw a horizontal line from the vertical "Temperature" axis to the data point representing the cold utility at 180 kW. We then find out how "high" we can shift this horizontal line upwards while maintaining the same cold utility duty. In this case, we realize that there are no higher temperatures that correspond to the same cold utility duty; therefore our **maximum cold utility temperature is unchanged at $T = 30\ °C$ (or shifted temperature $T^* = 35\ °C$).**

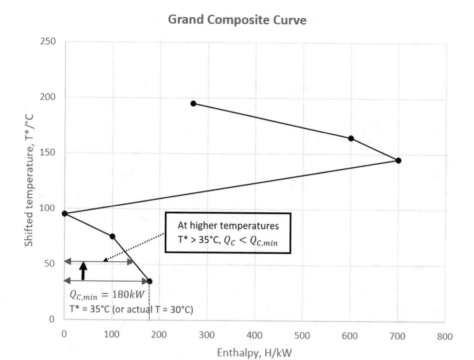

Problem 10

(a) **Explain what a grand composite curve is and briefly discuss its usefulness.**

(b) **Given the grand composite curve (above pinch) shown below and assuming that $\Delta T_{min} = 20$ K,**

1. **Annotate on the diagram the nose(s) of the curve and comment on its significance.**
2. **Determine the minimum hot utility duty, $Q_{H,min}$, for this system.**
3. **If only one hot utility was to be added, suggest its temperature and duty.**
4. **If two hot utilities were to be added at different temperatures, such that one is at the lowest possible temperature, suggest the temperatures and duties for the two hot utilities.**
5. **Compare between latent heat and sensible heat for utilities. Given that the hot utility in this problem utilizes latent heat effect for its heating duty, explain the difference between the setups in parts (3) and (4).**

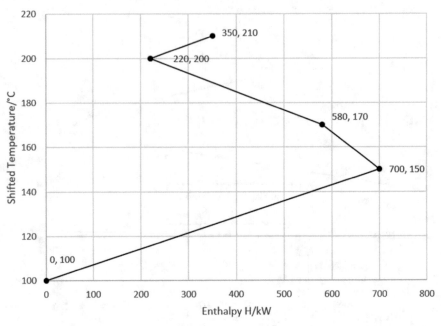

Solution 10

(a)

   The Grand Composite Curve (GCC) is a useful method of graphically depicting how heat energy flows from interval to interval. It expresses energy flow as a function of temperature across intervals. This type of representation is useful for various reasons including:

- Determining utility limits, e.g., maximum or minimum possible utility temperatures.
- Investigating different combinations of utilities that can be used to fulfill the heating and cooling loads of process streams, e.g., constant temperature vs variable temperature utility types.

   (b)(1)

   The "nose" of the GCC refers to the region whereby a vertical line encloses a region bounded on the other sides by the composite curve. This region represents an

inter-stream heat transfer region, in other words where there is heat transfer between process streams. As seen from the annotations in the graph below, noses occur where there are overlapping enthalpy values at different temperature sections of the composite curve.

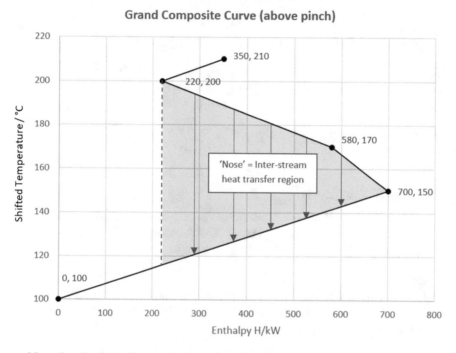

Grand Composite Curve (above pinch)

Note that the "nose" may also be referred to as "pocket." In general, we do not place external hot or cold utilities within the "nose" regions, since heat transfer in these regions can be achieved through process-to-process heat exchange which optimizes heat integration and energy recovery.

(2)

The minimum hot utility duty can be read off directly from the graph and the top of the curve. We find that for maximum energy recovery, $Q_{H,min} = 350$ **kW**.

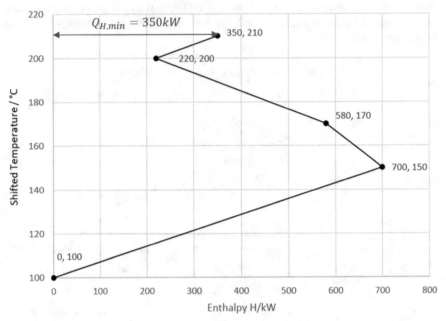

(3)

If the hot utility was only limited to one, then we can only supply the full duty of 350 kW at the shifted temperature of 210 °C, equivalent to an actual temperature of 220 °C as computed below.

$$T = T^* + \frac{\Delta T_{min}}{2} = 210 + \frac{20}{2} = 220^{\circ}\text{C}$$

(4)

If $Q_{H,\ min}$ can be split such that it is supplied via two hot utilities placed at different temperatures (and given that one of the temperatures is the lowest possible), then we can observe that graphically: the original horizontal line representing the total hot utility duty of 350 kW is now split into two parts, denoted $Q_{H,\ 1}$ and $Q_{H,\ 2}$.

$Q_{H,\ 1}$ will contribute **220 kW** of hot utility duty at the lowest possible temperature of $T^* = \mathbf{115.7}^{\circ}\text{C}$ (or actual temperature of 125.7 °C) while $Q_{H,\ 2}$ will contribute the remaining **130 kW** of hot utility duty at a higher temperature of $T^* = \mathbf{210}^{\circ}\text{C}$ (or actual temperature of 220 °C).

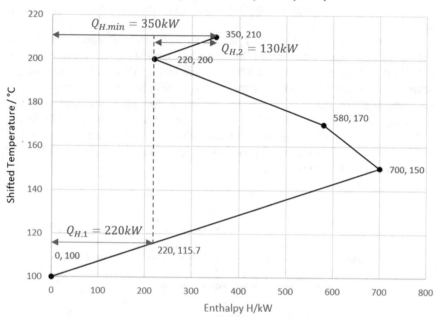

Grand Composite Curve (above pinch)

(5)
Latent heat differentiates from sensible heat such that there is **no temperature change** (or constant temperature) when latent heat is gained or lost during a phase change. On the other hand, sensible heat refers to heat gained or lost when there is a temperature change associated with the heating or cooling. If the hot utility was to be supplied using latent heat, the heat source could be provided by steam. High-temperature (or superheated) steam can be generated at higher pressures. Steam is widely used in most chemical plants as a hot utility due to the following advantages:

- High latent heat of condensation able to provide a large amount of heat energy per unit mass of steam at constant temperature.

  - This **high heat output of steam** is superior compared to other hot utilities such as hot oil or flue gas, which release sensible heat over a large temperature range. The efficiency of sensible heat transfer will depend on the heat capacities and flow rates of the fluids used since $Q = mc\Delta T$.
  - The condensation of steam has a desirably **large heat transfer coefficient**, which enables cost savings with **smaller heat exchangers needed** to be used (smaller heat transfer area needed).

- Steam is relatively **safe** to handle, as it is not toxic nor flammable. Moreover, any leaks can be visually detected.

Assuming that we use steam as the hot utility in our problem, the benefit of splitting the single hot utility of 350 kW at 220 °C (actual temperature) into two separate hot utilities, at (220 kW, 125.7 °C) and (130 kW, 220 °C), respectively lies in **cost savings** since part of the hot utility duty can be operated at a lower temperature.

Extreme operating conditions such as high temperatures and/or pressures result in substantial operating costs, which is undesirable, especially in the long run. Instead of using high-temperature steam at 220 °C to supply the **full** hot utility duty, we can achieve the same total hot utility duty that our system requires by using the high-temperature steam at 220 °C for only a fraction of the total utility duty, while the remaining utility duty can be fulfilled using lower temperature steam at 125.7 °C.

We can observe what happens to a utility when we change its temperature on a HEN grid display. Consider an arbitrary system with four process streams, where the grid display above the pinch is shown below.

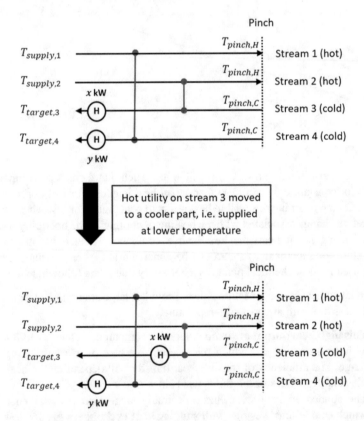

Notice that when we move a hot utility (for example, the one on cold stream 3) of a specified duty to a lower temperature, we are effectively moving it closer to the pinch (i.e., lowering its temperature as we move to the right). In the diagram below, $T_{pinch, H} - T_{pinch, C} = \Delta T_{min}$.

Problem 11

In a typical process plant, we can usually find various options for heating and cooling utilities, for example multiple steam levels, as well as a variety of coolant fluids with different heat capacities.

(a) **Briefly discuss the types of heating and cooling utilities available in industrial applications, in the context of constant and variable temperature utilities and the Grand Composite Curve (GCC).**

(b) **A student made the following statement about the GCC.**

"The GCC is useful graphical representation as it helps us determine if the generation of utilities is possible, such as raising steam via energy recovery."

Explain, using a suitable example(s), if you agree with this statement and discuss the broad principles in the optimization of multiple steam levels.

Solution 11

(a)

Heating and cooling utilities can be grouped into two main categories, *constant temperature* and *variable temperature* utilities.

Constant Temperature Utilities

- Constant temperature utilities remain at a fixed temperature and therefore appear as **horizontal lines** on the GCC.
- **Steam as a hot utility:**

  - One example is steam, which is a hot utility usually placed above the pinch (a net heat sink). Steam provides latent heat at constant temperature as it releases heat in the process of condensing into liquid water (phase change). The heat released serves as a heat source to the system; hence steam is a heating utility from the system's perspective. It follows that the constant temperature of steam is also the condensation or boiling/evaporation point.
  - Steam may be generated at various temperatures, whereby higher pressure steam is supplied at higher temperature and vice versa.

  Case 1: Single Steam Utility Above the Pinch, at High Pressure (HP)

Grand Composite Curve (above pinch)

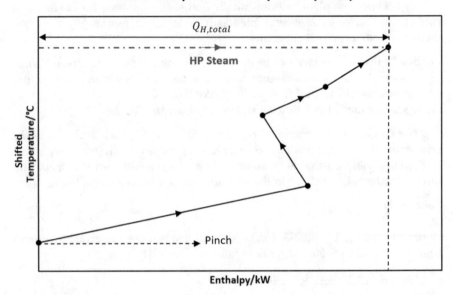

Case 2: Two Steam Utilities Above the Pinch, at (1) Medium-High Pressure (MHP) and (2) High Pressure (HP)

Grand Composite Curve (above pinch)

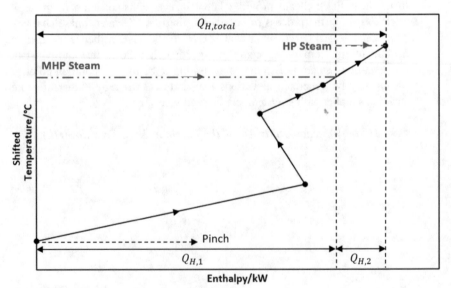

Key features of the GCC Plots:

- The gradients of the steam utility lines are zero, i.e., horizontal lines. This is due to the constant temperature nature of latent heat transfer.
- Single or multiple steam utilities may be used at different temperatures (hence pressures), as shown in cases 1 and 2. The total heating utility provided remains unchanged at $Q_{H, total}$. In case 2, $Q_{H, total} = Q_{H, 1} + Q_{H, 2}$.
- One advantage of case 1 is in the convenient use of just a single heating utility to supply the full heating duty. However, this comes at a "price" as it requires more energy to operate and maintain a higher pressure steam supply. The converse is true in case 2, where there will be a reduced operating cost with the splitting total heating duty into high and medium-high pressure steam where medium pressure steam is cheaper to operate than high pressure steam. However, this also comes at a "price" since greater capital expense is incurred for the installation of two hot utilities instead of one.

Variable Temperature Utilities

- Unlike constant temperature utilities, variable temperature utilities achieve their utility duties via sensible heat and a temperature change occurs as they go about their heating or cooling work. The exact amount of temperature change is determined by the utility fluid's heat capacity. Due to a non-zero temperature change, these utility lines are **sloped** on the GCC plot.
- **Coolant fluids as cold utilities:**

  - Variable temperature utilities are often cooling utilities (coolant fluids) below the pinch. This makes sense since we need something "cold enough" to take in heat from the system below the pinch (net heat source), and most coolant substances are in the liquid state at lower temperatures. Cooling water often suffices as a coolant, although we may also encounter other oils, and/or a combination of coolants used.
  - Generally, an ideal coolant fluid should have a relatively high specific heat capacity so that it can absorb substantial amounts of heat energy from the system without itself increasing in temperature excessively. It is desirable if a cooling utility remains cold as much as possible to fulfill its duty as a cold utility.
  - As the coolant fluid gains heat, it increases in temperature gradually. Therefore, a sloped line is seen on the GCC. The magnitude of the gradient is equivalent to the reciprocal of the coolant's heat capacity w as shown below:

    Magnitude of gradient of coolant line $= \frac{1}{w}$

Case 3: Two Coolant Fluids Below the Pinch, $w_1 < w_2$

Grand Composite Curve (below pinch)

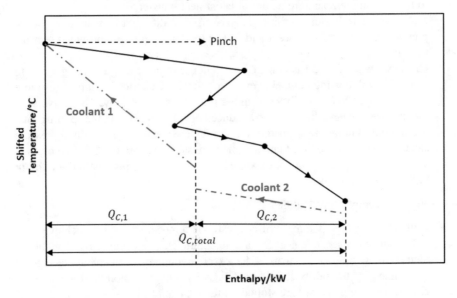

Enthalpy/kW

Key features of the GCC Plot:

- Since the slope of coolant 1's line is steeper (larger gradient) than that of coolant 2's, we deduce that the heat capacity of coolant 1 is smaller than that of 2, i.e., $w_1 < w_2$. This means that for every 1 K of temperature increase of the coolant fluid (it rises in temperature in the process of absorbing heat from the process streams), coolant 2 can absorb more heat energy (hence more effective cooling) than coolant 1.
- However, coolant 1 may still be included for practical reasons, for example it may be more stable (e.g., does not decompose or vaporize) at higher temperatures than coolant 2. Although coolant 2 has a greater heat capacity, its use may be restricted to lower temperature ranges.
- In the diagram above, we note that the total cooling utility duty is fulfilled by both coolants 1 and 2 via sensible heat, i.e., variable temperature utilities with lines of non-zero slopes.

$$Q_{C,\text{total}} = Q_{C,1} + Q_{C,2}$$

- Finally, we note the opposite directions along the vertical temperature axis in which the coolant and GCC lines move as process streams are cooled down below the pinch. The coolant line moves towards higher temperatures (upwards and to the left towards the pinch) while the GCC line is brought to lower temperatures. This obeys conservation of energy in the overall system. [Note that the horizontal enthalpy axis refers to enthalpy <u>of the process streams;</u> therefore, as the coolant lines reach higher temperatures along the arrows marked in the diagram above, the corresponding lower values of enthalpy (read off from the horizontal axis) represents a loss of heat energy of the process streams (not the coolant); separately the coolant gains that same amount of heat energy or enthalpy that is lost to it.]

Case 4: Two Coolant Fluids Below the Pinch, $w_2 < w_1$

Grand Composite Curve (below pinch)

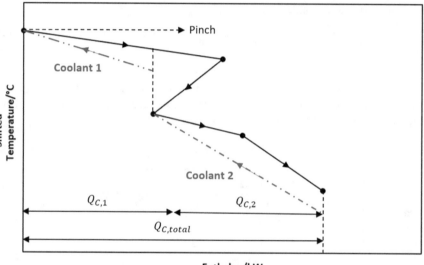

Key Features of the GCC Plot:

- Case 4 is similar to case 3 in that it also involves the combined use of two variable cooling utilities below the pinch. However, in this case, coolant 2 has a smaller heat capacity than coolant 1. In addition, coolant 1 operates at a higher temperature interval, which could mean that coolant 1 only remains in the liquid state at this higher temperature range and cannot be used at lower temperatures as it may solidify.

(b)

"The GCC is useful graphical representation as it helps us determine if the generation of utilities is possible, such as raising steam via energy recovery."

The student's statement is correct. It is possible to **raise steam using surplus heat** from process streams **below the pinch**. This means putting surplus heat to good use and hence the term "energy recovery." The GCC helps us visualize the temperatures and enthalpies of the process streams, thereby helping us determine suitable placements of utilities (e.g., constant and/or variable temperature utilities) to raise steam.

Optimization Principles for Utilities

- **Above the pinch:** Multiple steam levels are usually available. For example, we may have steam levels at high pressure (e.g., 200 °C shifted), medium-high pressure (e.g., 180 °C shifted), and low pressure (e.g., 140 °C shifted) respectively. In optimizing multiple steam levels, we generally <u>use hot utilities at the lowest possible temperature</u>.

- **Below the pinch:** Unlike steam levels above the pinch, we generally try to use cold utilities (e.g., steam raised below the pinch) at the <u>highest possible temperature</u>. In other words, we always try to raise steam from boiler feedwater at the highest possible temperature.

Let us now use the GCC plot to study a few possible scenarios that demonstrate the above points.

Above the Pinch—Three Possible Cases

Case 1: If we ignored the optimization principle, we would likely end up with Case 1 below, whereby the total heating duty of the process would be achieved using high pressure steam alone at 200 °C.

Grand Composite Curve

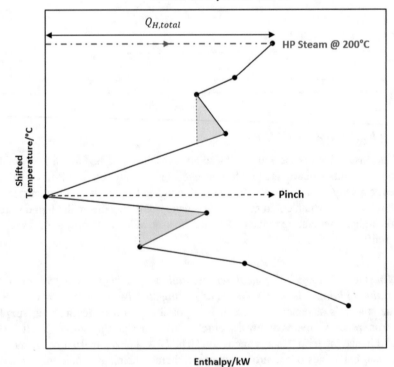

This arrangement is the least complex and incurs the **least capital cost since only one steam supply** needs to be installed. However, this arrangement bears **significant operating costs for maintaining steam supply at high pressure**. [Therein lies an inherent compromise between operating cost and capital cost.]

Case 2: In Case 2, we now split the total heating duty of the system between two steam supplies, one at high pressure and the other at medium-high pressure. Compared to Case 1, **Case 2 will incur a lower operating cost** but may bear a slightly higher capital cost to install two steam supplies instead of one.

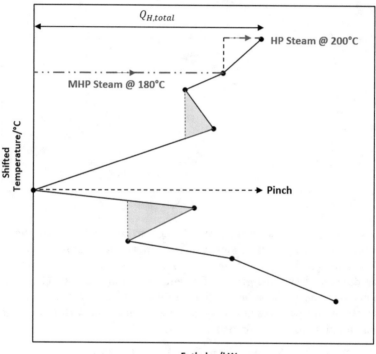

Grand Composite Curve

Case 3: Case 3 utilizes the most number of steam levels and hence best minimizes the operating cost. The total heating duty is fulfilled using steam at high pressure (200 °C), medium-high pressure (180 °C), and low pressure (140 °C).

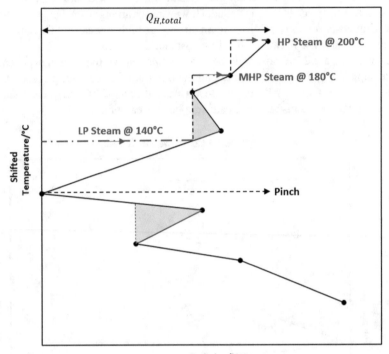

Grand Composite Curve

Enthalpy/kW

In the placement of the steam levels, we adopted the optimization principle by **using the greatest amount of steam at the lowest temperature available (i.e., 140 °C)**. After placing the first steam level at 140 °C, we continue fulfilling our heating duty by placing the other two steam levels 180 and 200 °C sequentially. [Note that the shaded region or the "nose" of the GCC represents inter-stream heat transfer where process streams satisfy each other's loads through heat integration; therefore utilities are not required at the "nose."]

Below the Pinch—Raising Steam from Boiler Feedwater

Below the pinch, we may optimize energy recovery by using some of the surplus heat from process streams to raise steam. Recall that the subsystem **below the pinch is a net heat source**, that is why we place cooling utilities below the pinch to accept heat that has been rejected by the process streams. Since the temperatures below the pinch are generally lower, even though we can raise steam below the pinch via energy recovery, the generated steam would likely be of lower pressures (and lower temperatures).

The process of raising steam is summarized as follows:

- Surplus heat from process streams is used to preheat boiler feedwater. Sensible heat is absorbed by the feedwater and its temperature increases from an initial temperature at the inlet to evaporation point (i.e., steam point for the low pressure steam to be generated).
- When evaporation point is reached, boiler feedwater in the liquid state converts into steam vapor, generating low pressure steam. Latent heat of vaporization is absorbed by boiler feedwater in this endothermic liquid-to-vapor phase change process, and this heat comes from surplus heat rejected by the process streams. In other words, we are coupling steam generation with the cooling of process streams.
- Steam generation provides two main benefits. It not only helps fulfill part of the total cooling utility duty below the pinch by recovering energy from surplus process heat but also produces a useful product, steam which can then be used for other parts of the process (e.g., the steam itself may be used for other heating duties).

Note that steam generation alone may not be sufficient to fulfill the total cooling utility duty below the pinch; therefore, we often require another coolant, e.g., cooling water, as a supplementary cooling utility to achieve the total cooling duty required below the pinch. Nonetheless, with the help of steam generation, we are able to minimize the cooling load on any additional coolants required, since only a fractional amount of the total cooling duty needs to be provided by the additional coolant. This reduced load thereby reduces operating costs on coolant utilities and also expands options for coolant fluids that can fulfill the smaller and more manageable cooling load.

Case Study: Combination of Constant Temperature and Variable Temperature Utilities for an Overall System

Let us consider an arbitrary case whereby the GCC plot of the process streams above and below the pinch is shown below. We may apply optimization principles both above and below the pinch to arrive at a utility configuration as indicated in the plot, which achieves energy recovery (via steam recovery) as well as minimization of operating costs of external utilities.

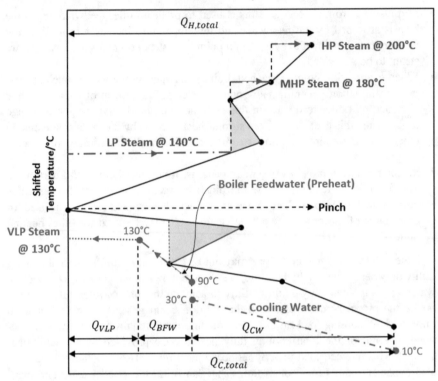

Grand Composite Curve

Below the Pinch—Combination of Utility Types

In this case study, we note that a combination of constant temperature and variable temperature utilities is used to fulfill the total cooling load $Q_{C, \text{total}}$ for the process streams below the pinch.

First Cold Utility—Cooling Water

- Working our way up the GCC (towards the pinch) from the low temperature end, we note that cooling water at 10 °C was first used to fulfill part of the cooling load below the pinch, Q_{CW}.
- The cooling water gains heat from the process streams and its temperature rises to a final temperature of 30 °C. The cooling water line is sloped since it is a variable temperature cooling utility that gains **sensible heat**, as opposed to latent heat.

Second Cold Utility—Preheat of Boiler Feedwater

- Boiler feedwater supplied at 90 °C was then used to absorb more heat from the process streams, an amount equivalent to Q_{BFW}. This is referred to as "preheat" in the graph since the heated feedwater will later be used to generate steam.
- Similar to cooling water, the boiler feedwater line is sloped since it gains **sensible heat** and experiences a temperature change from 90 to 130 °C.

Third Cold Utility—Vaporization of Boiler Feedwater

- When boiler feedwater reaches its evaporation point at 130 °C, vaporization takes place and this being an endothermic process, takes in heat from the process streams. The production of VLP steam from boiler feedwater at 130 °C is therefore said to be coupled to the cooling of process streams by an amount Q_{VLP}. [VLP denotes Very Low Pressure, to differentiate from higher steam levels such as LP, MHP, and HP steam above the pinch.]
- Unlike cooling water and boiler feedwater (90–130 °C), the line on the graph representing vaporization of feedwater at 130 °C is horizontal, indicating that it is a constant temperature utility that absorbs **latent heat**.

Application of Optimization Principles

- Recall from our optimization principles earlier that for steam generation below the pinch, it is optimal to raise steam from boiler feedwater at the **highest possible temperature**. This means that the horizontal VLP steam line should be placed as high as possible. This will allow us to generate the most useful steam (at highest possible temperature) and recover most energy while doing so, and at the same time removing heat from the process streams to fulfill their cooling loads below the pinch.
- It is also preferable that the horizontal VLP line extends over as wide a range of enthalpies as possible to generate more steam and hence recover more heat. This is acceptable as long as the **cooling utility lines (i.e., VLP, BFW, and CW) do not cross the GCC line**. Any crossovers would violate the law of thermodynamics, which states that heat must travel from hot to cold as shown below.

Grand Composite Curve

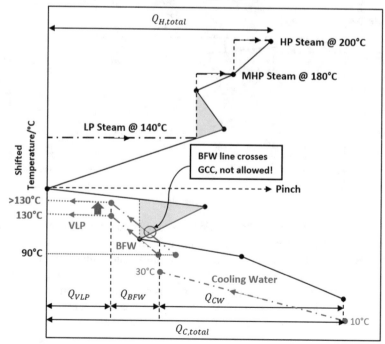

Enthalpy/kW

- As seen above, the highest possible temperature that steam can be raised from boiler feedwater in this example is 130 °C. If VLP steam was at any higher temperatures, the boiler feedwater line would begin to cross the GCC as marked by the red circle, which is not allowed.

  – In this particular example, 130 °C is the highest temperature of VLP steam that can be raised. However, it is possible to still produce steam levels higher than 130 °C if we adjust our operating conditions.
  – To correct the violation of crossing the GCC line, the boiler feedwater (BFW) line must have a steeper slope, or be supplied at a temperature higher than 90 °C.

    The slope of the BFW line is equivalent to the reciprocal of its flow rate heat capacity. Since flow rate can be altered, a lower flow rate will give us a steeper BFW line that would help avoid the crossover.

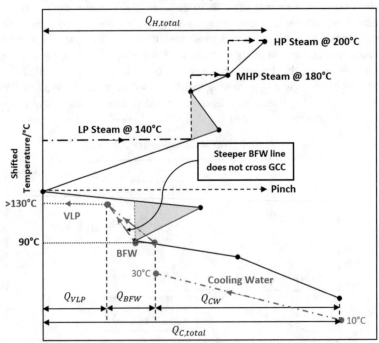

- The BFW could also be supplied at a temperature higher than 90 °C, such that it starts off at a temperature above the crossover point as shown below. However, this means that the cooling duty contributed by BFW preheat Q_{BFW} is also reduced. In order to still fulfill the total cooling duty of $Q_{C,total}$, we could increase Q_{CW} to compensate for this reduction. This means that cooling water at 10 °C will need to be heated to a higher temperature than before (>30 °C) to fulfill a larger cooling duty.

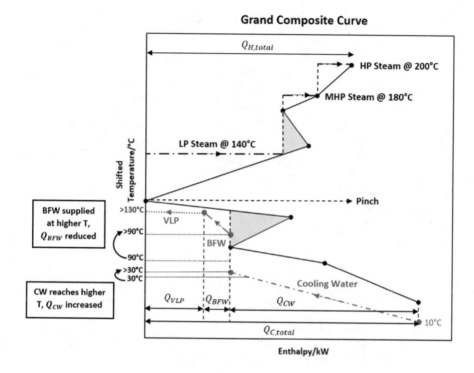

Grand Composite Curve

Problem 12
A heat exchanger network (HEN) for maximum energy recovery (MER) was designed for a process system and the corresponding Grand Composite Curve (GCC) is as shown below.

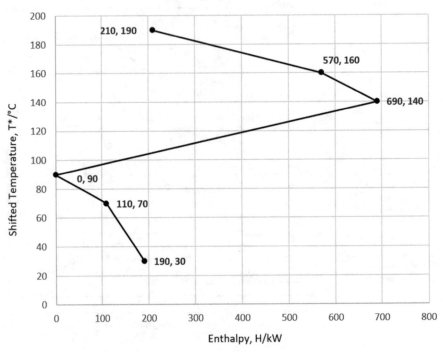

In the plot above, the shifted temperatures are based on a minimum approach temperature of 10 K at the pinch.

(a) **Explain what a pinch point is, and where it is found on a GCC plot. Determine the pinch temperatures and minimum utility duties for MER, and comment on how the GCC plot relates to an energy cascade.**

(b) **Given that a liquid hydrocarbon X available at 50 °C was to fulfill part of the system's cooling duty.**

1. **Comment on any condition(s) required for liquid X to be considered a suitable coolant for this system.**
2. **Determine the maximum cooling duty that can be achieved by liquid X, and hence the remaining utility cooling duty. Find the maximum flow rate heat capacity of liquid X that can achieve this maximum duty.**

Solution 12

(a)

The **pinch point splits the overall system into two decoupled parts; there is no heat transfer across the pinch:**

- Subsystem above the pinch is a <u>heat sink;</u> only heating utilities are present.
- Subsystem below the pinch is a <u>heat source;</u> only cooling utilities are present.

The pinch location occurs at the **point of closest approach** between the hot composite curve (sum of enthalpies of all hot streams over each temperature interval) and the cold composite curve (sum of enthalpies of all hot streams over each temperature interval). For example, in a system like the one below, the pinch point is as indicated.

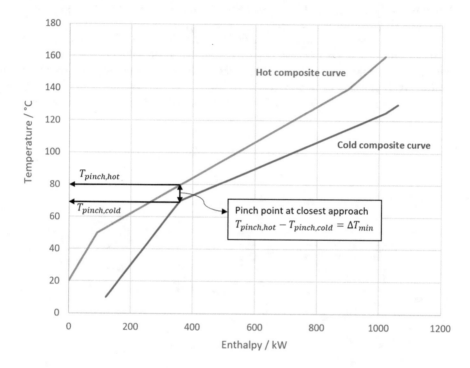

$$T_{\text{pinch,hot}} - T_{\text{pinch,cold}} = \Delta T_{\min}$$

It is given that the minimum approach temperature $\Delta T_{\min}$ in this problem is 10 K or $10\,^{\circ}\text{C}$, which when translated into shifted temperatures, $\Delta T_{\min,\text{ shifted}} = 0\,^{\circ}\text{C}$. This can be shown mathematically as follows at the pinch:

$$T_{\text{pinch,hot,shifted}} = T_{\text{pinch,hot}} - \frac{\Delta T_{\min}}{2}$$

$$T_{\text{pinch,cold,shifted}} = T_{\text{pinch,cold}} + \frac{\Delta T_{\min}}{2}$$

$$\Delta T_{\min,\text{shifted}} = T_{\text{pinch,hot,shifted}} - T_{\text{pinch,cold,shifted}}$$

$$\Delta T_{\min,\text{shifted}} = \left(T_{\text{pinch,hot}} - \frac{\Delta T_{\min}}{2}\right) - \left(T_{\text{pinch,cold}} + \frac{\Delta T_{\min}}{2}\right) = 0$$

By "shifting" temperatures, we reduce all hot stream temperatures by an amount of $\Delta T_{min}/2$ and raise all cold stream temperatures by the same amount $\Delta T_{min}/2$. If the pinch point occurs where the hot composite curve is ΔT_{min} higher than the cold composite curve, then shifting both curves' temperatures this way would graphically move the hot curve lower and cold curve higher such that they exactly meet at the pinch with $\Delta T_{min, \text{shifted}} = 0\,°C$. By using shifted temperatures, we simplify our pinch point to a **single point** where $T_{\text{pinch, hot, shifted}} = T_{\text{pinch, cold, shifted}}$.

The GCC shows the cumulative enthalpy of streams across shifted temperature intervals. In an energy cascade diagram, cumulative enthalpy is also cascaded enthalpy. Since there is no heat transfer across the pinch, **the value of enthalpy on the GCC plot has to be zero at the pinch**.

From the given GCC plot in the problem, we identify the pinch point where the enthalpy value is zero.

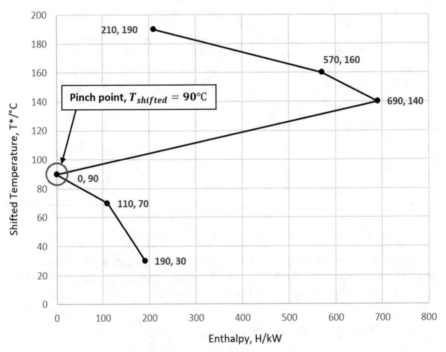

$T_{\text{pinch,hot,shifted}} - T_{\text{pinch,cold,shifted}} = 0, \text{ at the pinch}$

$T_{\text{pinch,hot,shifted}} = T_{\text{pinch,cold,shifted}} = 90\,°C$

We can find actual hot and cold pinch temperatures by undo-ing the shift as shown.

$$T_{\text{pinch,hot}} = T_{\text{pinch,hot,shifted}} + \frac{\Delta T_{\text{min}}}{2} = 90 + \frac{10}{2} = 95\,^{\circ}\text{C}$$

$$T_{\text{pinch,cold}} = T_{\text{pinch,cold,shifted}} - \frac{\Delta T_{\text{min}}}{2} = 90 - \frac{10}{2} = 85\,^{\circ}\text{C}$$

The minimum heating and cooling utilities can be found by finding the enthalpy values at the top and bottom of the GCC plot. $Q_{\text{H, min}} = 210$ kW and $Q_{\text{C, min}} = 190$ kW.

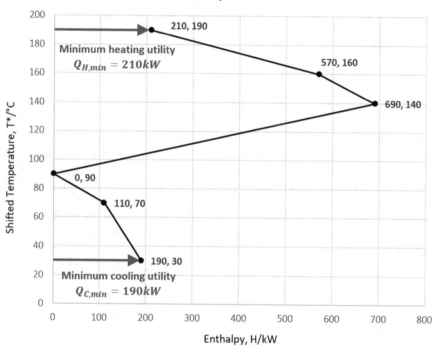

A GCC plot is a pictorial representation of the energy cascade. Heat energy or enthalpy first cascades down from the hot utility, then goes through a series of temperatures, before reaching the bottom-most receiving end where the cold utility is. Therefore, the enthalpies at the top and bottom of the Temperature-Enthalpy GCC plot give the minimum heating and cooling utilities respectively as shown below.

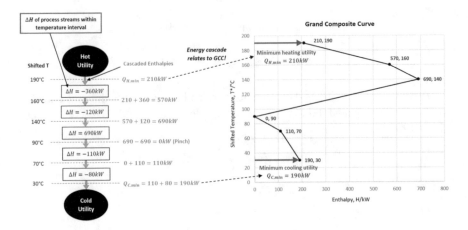

(b)(1)

A liquid coolant works by absorbing heat from the process stream(s) via a heat exchanger (HXer). This absorbed heat raises the temperature of the coolant fluid from a lower temperature at the inlet of the HXer to a higher temperature at its outlet. It is the **heat capacity of the liquid** that allows this "absorption" of heat which translates into its temperature rise. The greater the heat capacity of the fluid, the smaller the temperature rise will be. For a liquid coolant to function, it must therefore **remain in the liquid state throughout its operational range,** from inlet to outlet of the HXer. Therefore, the condition that hydrocarbon X must obey is to have a bubble point temperature greater than the maximum temperature below the pinch.

In this problem we note that for the section below the pinch on the GCC plot, the highest temperature reached would be at the pinch point, i.e., 90 °C (shifted) or 85 °C (actual). Therefore, the condition for hydrocarbon X is to have a **bubble point temperature exceeding 85 °C (actual).**

(2)

For a cooling utility such as a coolant fluid which gains **sensible heat**, its plot on the GCC diagram is a sloped line whereby the gradient of the line is equivalent to the reciprocal of coolant X's heat capacity denoted w_X.

$$\text{Slope} = \frac{1}{w_X}$$

The maximum cooling duty achievable by a coolant fluid corresponds to the fluid gaining the maximum possible amount of sensible heat. This corresponds to the case where the coolant line covers the greatest enthalpy range along its slope.

However, we need to note the following constraints on the coolant line:

- Coolant X's line must not cross the pinch into the section above the pinch. Cold utilities are only placed below the pinch and should not be found above the pinch.
- Any part of coolant X's line plot must not lie above the GCC curve (representing process streams). Since heat always travels from a hotter to colder region, the

hotter process stream(s) must lie above the cooler coolant fluid in order to allow heat transfer in the correct direction.

In this problem, we know that coolant X is supplied at 50 °C (or 55 °C shifted); therefore the starting or inlet temperature of X is fixed at 55 °C (shifted). To find the **maximum cooling duty** achievable, we need to position the coolant line such that it starts from 55 °C on the GCC diagram, moves upwards along a slope, and is able to **cover the widest enthalpy change when it reaches the pinch**. The entire path of the coolant line must also not cross the GCC curve or cross into the section above the pinch.

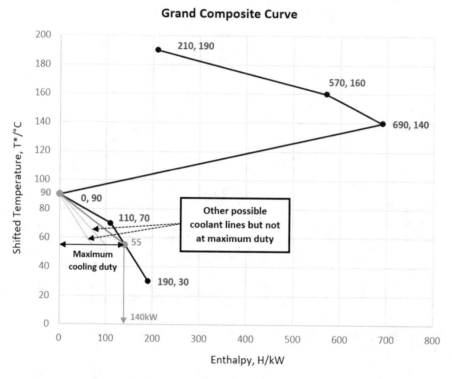

We find that the maximum duty that can be fulfilled by X is 140 kW, as shown above. Since the total cooling utility duty required (from part a) is 190 kW, the remaining utility is therefore 50 kW (=190 − 140).

Maximum Heat Capacity for X

Heat capacity of a coolant fluid has a reciprocal correlation with the gradient of its line on the GCC diagram. Therefore, a maximum heat capacity for X corresponds to the smallest gradient allowable, whereby the maximum duty of 140 kW can still be fulfilled.

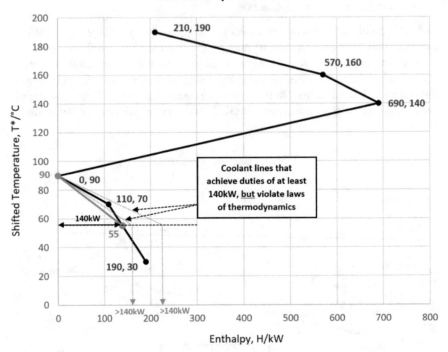

In the above plot, we observe that if we reduce the gradient of the coolant line (dotted lines above), i.e., increase its heat capacity, we would still be able to provide the 140 kW duty. However, doing so would violate laws of thermodynamics since the line would cross the GCC curve.

In order to achieve **both** a maximum cooling duty (140 kW) and have a maximum heat capacity (smallest gradient), the coolant line must be the one marked in orange bold line.

We can find the value of this maximum heat capacity by taking the reciprocal of the line's gradient.

$$\text{Gradient} = \frac{1}{w_X} = \frac{90 - 55}{140 - 0} = 0.25 \text{ K/kW}$$

$$w_X = \frac{1}{0.25} = 4 \text{ kW/K}$$

The maximum heat capacity is therefore 4 kW/K.

Chapter 4
Complex Hen Design Problems

Problem 13

Stream data for the section below the pinch are given as follows, whereby the minimum approach temperature $\Delta T_{min} = 20$ K and the pinch temperatures are 360 °C (cold) and 380 °C (hot).

Streams	Temperature range [°C]	Capacity flow rate [kW/K]
1 (hot)	380–180	10
2 (cold)	10–360	4
3 (cold)	40–340	2

(a) **One of your classmates commented that stream splitting is the only method that can be used to design a heat exchanger network (HEN) for this subsystem for maximum energy recovery. Discuss your classmate's comment and show how maximum energy recovery (MER) can still be achieved without splitting streams.**

(b) **Design a HEN for MER using the stream splitting approach.**

(c) **Compare your results in parts (a) and (b) using Euler's Network theorem, $U = N + L - S$. Highlight any key differences.**

Solution 13

(a)

Given the three streams below the pinch, we note that **stream 1 is the only hot stream** while **streams 2 and 3 are cold streams**.

Below the pinch, the following conditions have to be fulfilled.

- Number of hot streams greater or equal to the number of cold streams, $N_H \geq N_C$
- Only cooling utility present, Q_C

In addition, for stream matches **at the pinch (cold side)**,

© The Editor(s) (if applicable) and The Author(s), under exclusive license to Springer Nature Switzerland AG 2021
X. W. Ng, *Concise Guide to Heat Exchanger Network Design*,
https://doi.org/10.1007/978-3-030-53498-1_4

- Heat capacity of hot stream has to be greater or equal to that of cold stream, $W_H \geq W_C$

Looking at our stream table, we note that the condition $W_H \geq W_C$ at the pinch can be fulfilled, since the heat capacity of hot stream 1 is greater than that of both cold streams 2 and 3.

However, we **are unable to fulfill the condition** $N_H \geq N_C$, since $N_H = 1$ and $N_C = 2$. Stream splitting is useful in solving this as it helps us create two hot streams from the single hot stream. We will then be able to fulfill the equality sign of the condition, $N_H \geq N_C$. Note that when an original stream is split, its heat capacity value is also split into two smaller values. Therefore, using this method, we should be reminded to check that the new flow rate heat capacities of the split streams still fulfill the condition $W_H \geq W_C$ for matches at the pinch.

While the classmate's comment about stream splitting is correct, it is **not the only method** to design a HEN to that achieves MER for this system.

Cyclic Matching
We can still design a HEN for MER by introducing "loops" into the system. Recall Euler's network theorem as follows, whereby U refers to the number of heat exchanger (HXer) units, N refers to the total number of streams and utilities, L represents the number of loops, and S represents the number of individual subsystems.

$$U = N + L - S$$

One advantage of the stream splitting method is that it avoids creating unnecessary loops. The more loops we have, the more HXer units we will need. An optimal HEN design is one whereby we can achieve maximum energy recovery with a minimum number of HXers, which means a minimum number of loops (ideally $L = 0$). When we are able to "break" a loop (i.e., reduce the value of L), we are also removing a HXer unit (U decreases as L decreases).

This method of cyclic matching does not split streams, but it introduces loops to the system and hence uses more HXers in its HEN design than the stream splitting method.

Method
Let us first calculate the heating and cooling loads of the respective process streams and indicate them alongside our grid display as shown below.

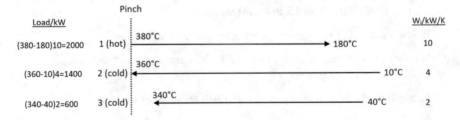

First Match (HXer 1)
Given that $\Delta T_{min} = 20$ K at the pinch, we deduce here that the first match at the pinch has to be between stream 1 at 380 °C (hot pinch) and stream 2 at 360 °C (cold

pinch) since this is the only pair that fulfills ΔT_{min}. Matching streams 1 and 3 would have exceeded the given ΔT_{min}, since $\Delta T_{min} = 380 - 340 = 40°C > 20$ K.

This match also fulfills the other required condition at the pinch (cold side) $W_H \geq W_C$, since $W_1(=10) > W_2(=4)$.

We may now update our grid display with the first stream match as shown below.

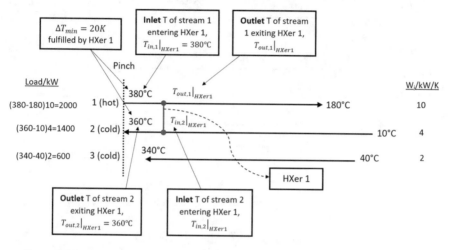

Second Match (HXer 2)

The next logical match in HXer 2 is between streams 1 and 3 since we have just used stream 2 as the cold stream in the HXer 1. Stream 1 is used again as the hot stream in HXer 2 since it is the only available hot stream.

We observe that the outlet temperature of stream 1 from HXer 1 will become the inlet temperature of stream 1 entering HXer 2. Similarly, the exit temperature of stream 3 from HXer 2, $T_{out,\,3}|_{HXer2}$, has to be equivalent to $340°C$, the final target temperature of stream 3 given in the problem.

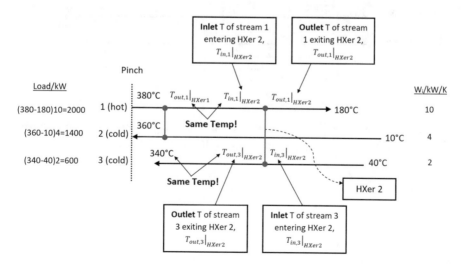

Since $\Delta T_{min} = 20$ K, we know that the lowest temperature that stream 1 can have coming out of HXer 1 and entering HXer 2 is $360\,^\circ$C ($=340 + 20$). Any lower would violate the ΔT_{min} condition for HXer 2 since ΔT_{min} would then be less than 20 K.

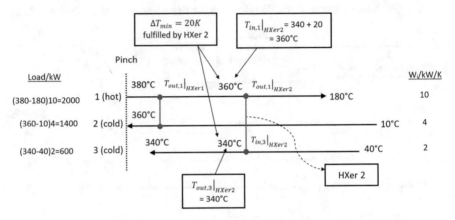

Now that we know $T_{in,\,1}|_{HXer2} = 360\,^\circ$C, we can relate back to $T_{out,\,1}|_{HXer1}$ since

$$T_{out,1}\big|_{HXer1} = T_{in,1}\big|_{HXer2} = 360\,^\circ C$$

Knowing the inlet ($380\,^\circ$C) and outlet ($360\,^\circ$C) temperatures of stream 1 for HXer 1 as well as the heat capacity of stream 1 allows us to compute the duty for HXer 1, $Q_1 = (380 - 360)W_1 = (20)10 = 200$ kW.

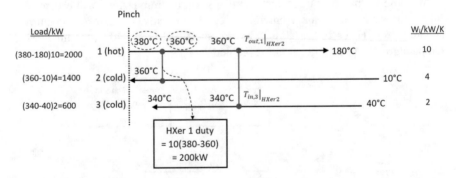

Using the duty of HXer 1, we can calculate the inlet temperature of cold stream 2 entering HXer 1. $T_{in,2}|_{HXer1} = 360 - \frac{200}{W_2} = 360 - \frac{200}{4} = 310\,^\circ$C.

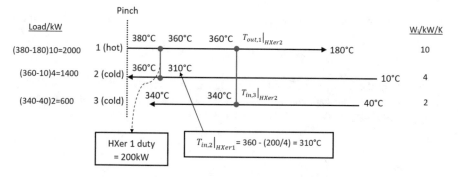

Third Match (HXer 3)

Now that we have completed the second match, we deduce our next match to be between streams 1 and 2 again since we have just used stream 3 as the cold stream for the previous match (i.e., second match). At this point, we may observe a "cyclic matching" pattern where we alternate between cold streams 2 and 3 in placing further matches with stream 1. We will continue placing matches until we completely fulfill the heating and cooling loads of all our process streams.

For the third match, we note that stream 2 must have a temperature of $310\,^{\circ}C$ exiting from HXer 3. $T_{out,\,2}|_{HXer3} = 310\,^{\circ}C$.

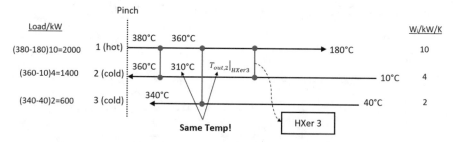

Since $\Delta T_{min} = 20$ K, we know that the lowest temperature that stream 1 can have coming out of HXer 2 and entering HXer 3 is 330 °C (=310 + 20). Any lower would violate the ΔT_{min} condition for HXer 3 since ΔT_{min} would then be less than 20 K.

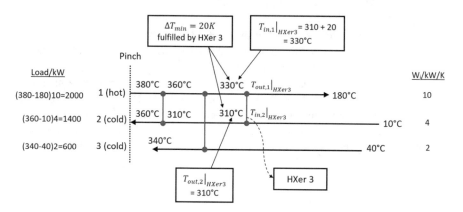

Now that we know $T_{\text{in, 1}}|_{\text{HXer3}} = 330°\,C$, we can relate back to $T_{\text{out, 1}}|_{\text{HXer2}}$ since

$$T_{\text{out,1}}|_{\text{HXer2}} = T_{\text{in,1}}|_{\text{HXer3}} = 330°\,C$$

Knowing the inlet $(360°\,C)$ and outlet $(330°\,C)$ temperatures of stream 1 for HXer 2 as well as the heat capacity of stream 1 allows us to compute the duty for HXer 2, $Q_2 = (360 - 330)W_1 = (30)10 = 300$ kW.

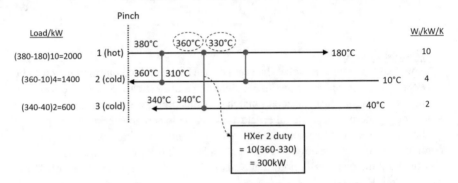

Using the duty of HXer 2, we can calculate the inlet temperature of cold stream 3 entering HXer 2. $T_{\text{in,3}}|_{\text{HXer2}} = 340 - \frac{300}{W_3} = 340 - \frac{300}{2} = 190°\,C$.

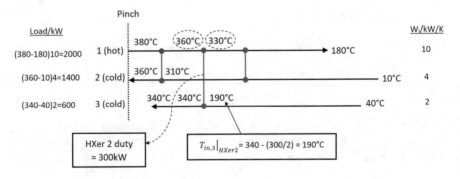

Fourth Match (HXer 4)

Next, we proceed to place the fourth match or HXer 4.

We know that HXer 4 will be a match between streams 1 and 3 following the cyclic matching pattern earlier mentioned.

Similar to what was done for previous HXers, we note again that the temperature of stream 1 at the inlet of HXer 4 must be 210 °C (=190 + 20), so that it is $\Delta T_{\text{min}} = 20$ K higher than the outlet temperature of stream 3 exiting from HXer 4, $T_{\text{out, 3}}|_{\text{HXer4}}$. We also know that $T_{\text{out, 3}}|_{\text{HXer4}}$ is just 190 °C since it is a continuous stream leading to the inlet of HXer 2 at the same temperature of 190 °C.

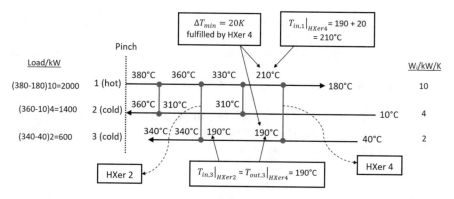

Now that the inlet (330°C) and outlet (210°C) temperatures of HXer 3 for stream 1 are also known, we can compute the duty for HXer 3, $Q_3 = (330 - 210) W_1 = (120)10 = 1200$ kW.

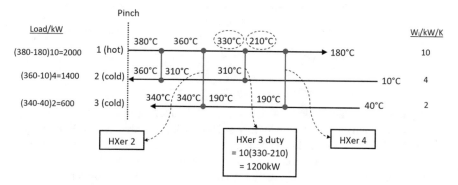

Using the duty of HXer 3, we can calculate the inlet temperature of cold stream 2 entering HXer 3. $T_{in,2}|_{HXer3} = 310 - \frac{1200}{W_2} = 310 - \frac{1200}{4} = 10°C$. This means that after placing the third HXer, we have already arrived at the supply temperature of stream 2, $T_{supply, 2} = 10°C$, and fulfilled its load completely. Therefore stream 2 does not need to participate in any further matches after the third match.

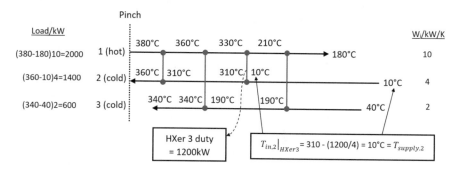

As for streams 1 and 3, we check if their remaining loads are also completely fulfilled with the fourth match.

Remaining load for hot stream 1 $=(210°C - T_{target, 1})W_1 = (210°C - 180°C)$ $10 = 300$ kW.

Remaining load for cold stream 3 $=(190°C - T_{supply, 3})W_3 = (190°C - 40°C)$ $2 = 300$ kW

If the duty for HXer 4 was 300 kW, then all three streams' heating and cooling requirements would be completely fulfilled without the use of any external utilities.

The final grid display is shown below.

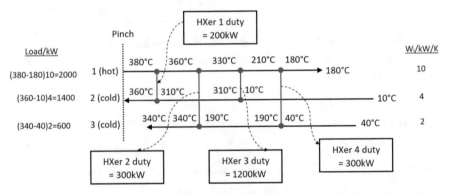

As we can see, the full load of stream 1 (i.e., 2000 kW) can be fulfilled by HXers 1, 3, and 4, which have a combined duty of 200 + 1200 + 300 = 2000 kW. As for the full load of stream 2 (i.e., 1400 kW), it can be fulfilled by HXers 1 and 3, with a combined duty of 200 + 1200 = 1400 kW. Finally, for stream 3, its full load of 600 kW can be fulfilled completely by HXers 2 and 4 with a combined duty of 300 + 300 = 600 kW.

In this system (below the pinch) where no external utilities are required, the corresponding GCC curve (using shifted temperatures) will have a shape as shown below whereby the GCC plot starts at the pinch and goes back to an enthalpy of zero.

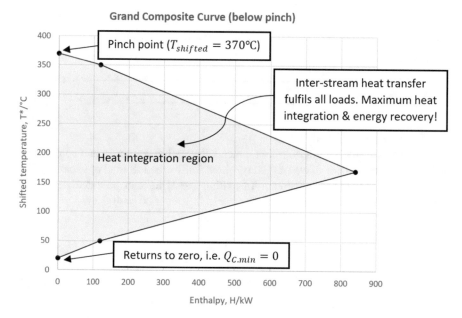

Grand Composite Curve (below pinch)

Pinch point ($T_{shifted} = 370°C$)

Inter-stream heat transfer fulfils all loads. Maximum heat integration & energy recovery!

Heat integration region

Returns to zero, i.e. $Q_{c.min} = 0$

Shifted temperature, $T^*/°C$

Enthalpy, H/kW

Both the stream splitting method and the cyclic matching method achieve maximum energy recovery without the use of external utilities. Both designs allow all streams to achieve their respective target temperatures via process-to-process heat transfer alone.

(b)

Recall that below the pinch, the condition for the number of hot and cold streams $N_H \geq N_C$ has to be fulfilled. Furthermore, matches at the pinch are to fulfill the condition for flow rate heat capacities $W_H \geq W_C$.

With stream splitting, we obtain two hot streams which we shall denote streams 1a and 1b here. This allows us to fulfill the equality sign of the condition, $N_H \geq N_C$ since $N_H = N_C = 2$. When stream 1 is split, its heat capacity $W_{H,\,1}$ is also split into two such that $W_{H,\,1} = 10 = W_{H,\,1a} + W_{H,\,1b}$.

In assigning specific values for $W_{H,\,1a}$ and $W_{H,\,1b}$, we should be mindful that matches at the pinch (cold side) are to obey $W_H \geq W_C$. As a general practice for all matches, we should also try to fulfill the full load of one or more streams as far as possible.

Consider the scenario where we split stream 1 into two streams of equivalent heat capacity, i.e., $W_{H,\,1a} = W_{H,\,1a} = 5$ kW/K as shown below with the respective loads of the four streams indicated.

Load/kW   W_i/kW/K

(380-180)5=1000   1a (hot)   380°C ————————————————→ 180°C   5

(380-180)5=1000   1b (hot)   380°C ————————————————→ 180°C   5

(360-10)4=1400   2 (cold)   360°C ←———————————————— 10°C   4

(340-40)2=600   3 (cold)   340°C ←———————————————— 40°C   2

Match Between 1a and 3

Let us arbitrarily consider our first match between hot stream 1a and cold stream 3. This gives us a HXer of 600 kW duty which fulfills the full load of stream 3. The intermediate temperature of stream 1a exiting the HXer can be calculated to be 260 °C, while the remaining load of stream 1a is 1000–600 = 400 kW.

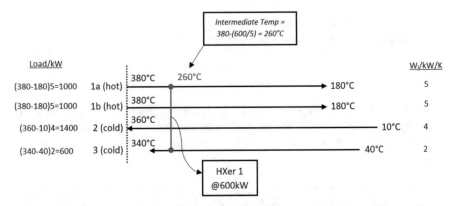

Note that this is not a match at the pinch, since stream 3 is not at pinch temperature of 360 °C; therefore the condition $W_H \geq W_C$ need not be fulfilled. A match at the pinch requires both streams to be at pinch temperatures.

Match Between 1b and 2

For the next match, we can pair up both remaining unmatched streams 1b and 2. This gives us a HXer of 1000 kW duty which fulfills the full load of stream 1b. Stream 2 leaves the HXer at an intermediate temperature of 110 °C, while the remaining load of stream 2 is 1400–1000 = 400 kW.

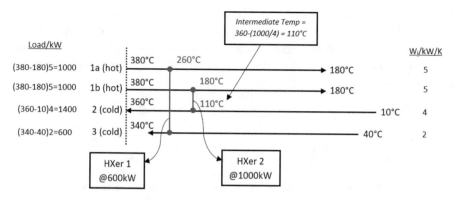

Note that this is a match at the pinch since both streams are at pinch temperatures; therefore we will need to check if this match fulfills the condition $W_H \geq W_C$, which is the case since $W_H(=5) \geq W_C(=4)$.

<u>Match Between 1a and 2</u>

It is easy to observe that after placing the first two matches, the only remaining streams with unfulfilled loads are hot stream 1a and cold stream 2. Therefore, the next logical match would be between these streams, giving us a third HXer of 400 kW duty which fulfills the remaining loads of both streams. Note that this match is not a match at the pinch since both streams are not at pinch temperatures. Hence, it is not necessary to obey $W_H \geq W_C$ for this match.

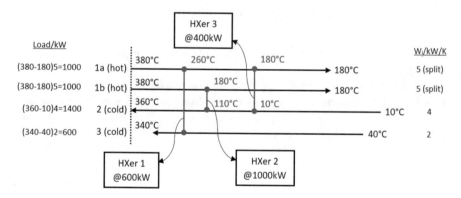

With the placement of the third HXer, we have also completed our HEN design for this system. Combining all results, we arrive at our final grid display above. Our HEN design for maximum energy recovery with stream splitting consists of 3 HXer units and requires no external utilities.

(c)

As mentioned in part (a), we expect to have more HXer units with the cyclic matching method as compared to the stream splitting method since the former introduces loops. Let us explore this using Euler's network theorem, $U = N + L - S$. For the **cyclic matching approach**, we have:

- $N = 3$ since we have three streams and no external cooling utilities.
- $L = 2$ since we observe two loops in the grid display as outlined in dotted lines below, which result from matching the same hot and cold streams multiple times. Note that a loop is defined as a closed path formed by connecting stream matches and/or utilities in a complete loop.

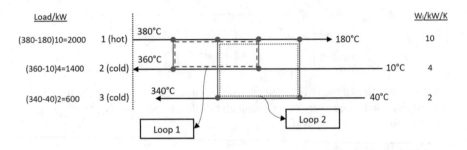

- $S = 1$, since we have one subsystem (i.e., independent section below the pinch).
- Euler's equation gives us a value of $U = 4$ as shown below, corresponding to 4 HXer units used.

$$U_{cyclic} = 3 + 2 - 1 = 4$$

For the **stream splitting approach** as worked out in part (b), we have the following.

- $N = 4$ since we have four streams after we split stream 1 into two, and no external cooling utilities.
- $L = 0$ since we observe no loops from the grid display.
- $S = 1$ since we have one subsystem (i.e., independent section below the pinch).
- Therefore, Euler's equation gives us a value of $U = 3$ as shown below, corresponding to 3 HXer units used.

$$U_{stream\ split} = 4 + 0 - 1 = 3$$

The results show that the stream splitting method used one less HXer than the cyclic matching method, despite the addition of a stream. This is due to the breaking of "loops" present in the cyclic matching HEN design.

Problem 14

Euler's Network Theorem as shown in the following expression is often applied to heat exchanger network (HEN) designs.

$$U = N + L - S$$

1. **Discuss what the symbols in the equation above mean and how they may be determined in HEN design.**
2. **Given the following HEN design data,**

Stream	Flow rate heat capacity W, [kW/K]	Supply temperature, [°C]	Target temperature, [°C]	Full load, [kW]
1 (hot)	6	175	70	?
2 (hot)	4	135	40	?
3 (cold)	5	40	140	?
4 (cold)	5	90	180	?

1. **Calculate the full loads for each stream and complete the table above.**
2. **Given the following stream matches and utility duties, illustrate the HEN design on a grid display.**
   Above pinch:

   • **Match between streams 1 and 4—450 kW of heat transfer**
   • **Match between streams 2 and 3—220 kW of heat transfer**
   • **Hot utility on stream 3—50 kW**

   Below pinch:

   • **Match between streams 1 and 3—180 kW of heat transfer**
   • **Match between streams 2 and 3—50 kW of heat transfer**
   • **Cold utility on stream 2—110 kW**

3. **Euler's network can be diagrammatically represented using graphs comprising nodes and edges. Illustrate your HEN design from part (2) in a graph diagram and comment on the presence of any loops. Use Euler's equation to confirm the number of loops present.**

Solution 14
(a)

Euler's network equation is shown below and the meaning of the symbols are explained as follows.

$$U = N + L - S$$

U—**Number of heat exchanger (HXer) units**

• This includes the number of **stream matches and utilities**. In the optimization of HEN designs, it is common to try to achieve a minimum value for U or U_{min} as this means satisfying heating and cooling needs of our system without the excessive use of HXer units.
• It follows that U_{min} can be found by imposing the condition of having no loops in our HEN design ($L = 0$ in Euler's equation).

- If we have hot (or cold) **utilities placed on separate streams**, we should **count them separately in deriving the value of** U, i.e., if we had one hot utility on one stream and another hot utility on another stream, then U from utilities should be 2. [Note that this is different from how we consider utilities in the computation of N—see below.]

N—Number of streams and utilities

- This value can be obtained simply by summing up the total number of hot and cold streams, as well as external hot and cold utilities.
- When counting utilities, it is worth noting that **if we had multiple cold utilities placed on separate streams**, the contribution of cold utilities to the value of N is 1 (and not 2). Similarly, if we had multiple hot utilities placed on separate streams, the hot utility count towards N is 1. In other words, all cold utilities are counted collectively as one, and the same is done for hot utilities. [Note that this is different from how we consider utilities in the computation of U—see above.]

L—Number of loops

- A loop is also a closed path of heat exchange. This path is formed by going along the heat exchange path formed using HXers and/or utilities in a closed loop, hence the path returns to its starting point.
- Examples of loops:

    Example 1: A loop involving only stream matches

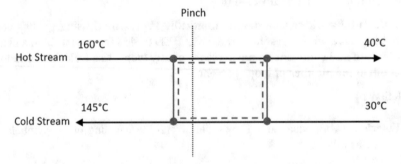

Example 2: A loop involving stream matches and utilities

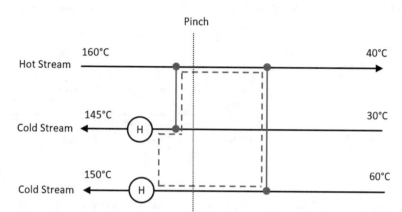

- A loop is also sometimes referred to as a "heat load loop" because HXers and/or utilities connected in a loop can vary their heat loads (i.e., duties) **without** affecting the target temperatures of the streams. Note that the number of loops L has implications on the number of HXer units U. The more loops there are, the more HXer units would be required.

S—Number of subsystems

- S represents a **subsystem or subnetwork**, which is a smaller secondary HEN which **satisfies its own heat energy balance internally** within this subsystem.

(b)(1)
The load of each stream can be calculated as follows.

- For hot streams: Cooling load = Heat capacity × (Supply temperature − Target temperature)
- For cold streams: Heating load = Heat capacity × (Target temperature − Supply temperature).

Stream	Flow rate heat capacity W, [kW/K]	Supply temperature, [°C]	Target temperature, [°C]	Full load, [kW]
1 (hot)	6	175	70	6(175–70) = **630**
2 (hot)	4	135	40	4(135–40) = **380**
3 (cold)	5	40	140	5(140–40) = **500**
4 (cold)	5	90	180	5(180–90) = **450**

(2)

The grid display showing the streams, stream matches, and utilities is as follows.

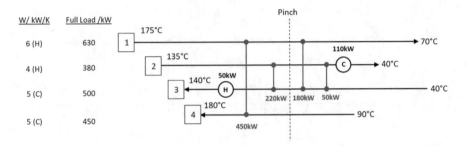

(3)

We can represent our HEN design using a graph diagram whereby streams and utilities are indicated as "nodes" in circles while heat transfers are indicated as "edges" in arrows. The respective loads of the streams as well as the utility duties are indicated by values within the circles, while the heat transfer duties are indicated as values alongside the arrows.

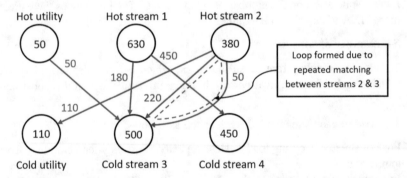

We notice from the graph diagram that there is a loop formed from the repeated matching between streams 2 and 3. This also means that the total number of HXers in this HEN design is not minimized as we can further reduce the number of HXers used by "breaking" (or removing) the loop.

Using Euler's network equation $U = N + L - S$, we can determine the number of loops and confirm what we observe from the graph diagram (i.e., $L = 1$ for one loop).

- Since we have four stream matches, one hot utility, and one cold utility, the number of HXers, $U = 6$.
- Next, we find the value of $N = 4 + 1 + 1 = 6$ since we have four process streams, one hot utility, and one cold utility.
- Finally, we have $S = 1$ for the overall system.

We can now infer the number of loops in our system by using Euler's network equation.

$$U = N + L - S$$
$$6 = 6 + L - 1$$
$$L = 1$$

Euler's network theorem confirms that one loop is present, similar to what we observed from the graph diagram. We may also notice this loop in our grid display from part (2).

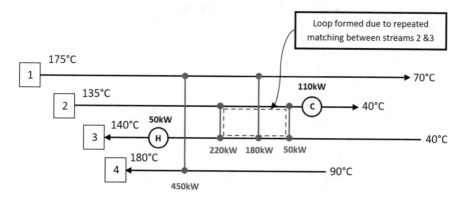

Problem 15

A student commented that based on Euler's network equation, we can reduce the number of heat exchanger (HXer) units required by either reducing the number of loops or increasing the number of subsystems. Show how the latter is possible using the following graph diagram and determine the number of HXer units U required before and after the change is made.

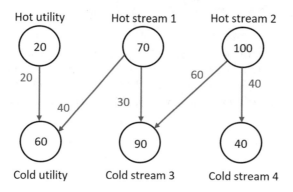

[Note: streams and utilities are indicated as "nodes" in circles while heat transfers are indicated as "edges" in arrows. The numbers represent stream loads and heat transfer duties in kW.]

Solution 15

The student is right since Euler's network equation is represented by the following equation; therefore mathematically we observe that we can reduce the value of U by decreasing L or increasing S.

$$U = N + L - S$$

We can demonstrate how increasing the number of subsystems can also reduce the number of HXers used, U. From the graph diagram given in the problem, we observe four process streams, one hot utility, and one cold utility. Also, there are no observable loops; therefore $L = 0$.

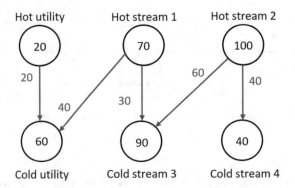

Using Euler's network equation, the value of U may be computed.

$$U = N + L - S$$
$$U = (4 + 1 + 1) + 0 - 1$$
$$U = 5$$

Since there are no loops in this HEN design, there is no scope to reduce the number of HXers used (i.e., $U = 5$) by reducing the number of loops, L. However, we can still reduce U by re-designing our HEN network such that we increase the value of S (number of subsystems). One possible re-design is shown in the graph diagram below.

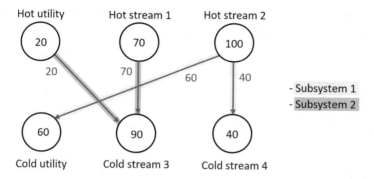

Hot utility — Hot stream 1 — Hot stream 2

20 — 70 — 100

20 — 70 — 60 — 40

- Subsystem 1
- Subsystem 2

60 — 90 — 40

Cold utility — Cold stream 3 — Cold stream 4

In this design, we have the same number of streams and utilities as before; therefore $N = 6$. And as before, there are no loops observed, $L = 0$. However, we now have **two subsystems**, separately marked in yellow and green. Within each subsystem, the heat source(s) (where arrows point out) and heat sink(s) (where arrows point towards) have exactly matching thermal duties; hence enthalpy balances internally within each subsystem and they are thermally independent.

With this design, $S = 2$, and the new value of U is 4, which is reduced from 5 from the original HEN design.

$$U = N + L - S$$
$$U = (4 + 1 + 1) + 0 - 2$$
$$U = 4 < 5$$

Problem 16
An engineer has designed a heat exchanger network (HEN) for maximum energy recovery for a process system consisting of two hot streams, two cold streams, as well as an external heating utility on stream 3 and an external cooling utility on stream 2. It is assumed that the minimum approach temperature is 20 K at the pinch. The grid diagram below shows his HEN design, with some information accidentally removed.

Flowrate heat capacity / kW/K	Thermal Load (above pinch) /kW		Pinch			Thermal Load (below pinch) /kW
6 (Hot)	?	1 180°C	? °C		70°C	?
3 (Hot)	?	2 160°C	? °C	C 40°C	? kW	?
4 (Cold)	?	3 145°C H	90°C	180kW 60kW 30°C		?
8 (Cold)	?	4 150°C 40kW	180kW 90°C			
		? kW				

(a) **Fill in the blanks in the grid diagram above to complete his HEN design.**
(b) **Fill in the blanks in the table below using information from the grid diagram (answer to part (a)) as well as Euler's Network Theorem**

$U = N + L - S$ for **HEN design. Comment briefly on how Euler's results below relate to the HEN design in part (a).**

Variables	Above the pinch	Below the pinch
U	3	3
N	?	4
L	0	?
S	?	?

Solution 16

(a)

This problem is easiest to solve by identifying the particular missing information that will help us determine the rest of the missing information more easily, i.e., the hot pinch temperature.

Given that the minimum approach temperature or $\Delta T_{min} = 20\ K = 20\,^{\circ}C$, and that the cold pinch temperature is $90\,^{\circ}C$ as observed from the grid display, we can find the hot pinch temperature which is missing for streams 1 and 2.

$$T_{pinch,hot} = 90\,^{\circ}C + \frac{\Delta T_{min}}{2}$$

$$T_{pinch,hot} = 90 + \frac{20}{2} = 100\,^{\circ}C$$

We can now update the grid display with this result.

Next, we observe that since the flow rate heat capacities for all streams are given, we can determine their thermal loads both above and below the pinch.

Stream	Flow rate heat capacity [kW/K]	Thermal load (above pinch) [kW]	Thermal load (below pinch) [kW]
1	6	$6(180 - 100) = 480$	$6(100 - 70) = 180$
2	3	$3(160 - 100) = 180$	$3(100 - 40) = 180$
3	4	$4(145 - 90) = 220$	$4(90 - 30) = 240$
4	8	$8(150 - 90) = 480$	$-$

Using the known thermal loads, we can now back-calculate the missing duties for the heat exchangers (HXers) between streams, as well as external utilities.

Above the Pinch

The HXer duty for the match between streams 1 and 4 is missing. Since stream 1 is only matched with stream 4 and stream 4 is only matched with stream 1, and both streams 1 and 4 do not have external utilities placed on them, the full thermal loads of streams 1 and 4 above the pinch should completely satisfy each other in this match.

We have found in the table above that stream 1's thermal load (above the pinch) is indeed equivalent to stream 4's thermal load (above the pinch), i.e., 480 kW. It therefore follows that the HXer duty between streams 1 and 4 above the pinch would be 480 kW.

Below the Pinch

Next, we determine the cold utility duty placed on stream 2. Again, we start by finding the thermal load for stream 2 below the pinch, which we found earlier to be 180 kW. We next observe that below the pinch, other than the cooling utility on stream 2, there is a stream match between streams 2 and 3 at a heat exchange duty of 60 kW. Therefore, we can back-calculate the cooling utility duty required for heat balance.

$$Q_{c,\,min} = 180 - 60 = 120 \text{ kW}$$

We can now update all our results in the grid diagram and obtain the complete HEN design.

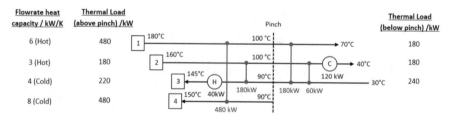

(b)

Finding the value of U

From Euler's network equation, the value of U corresponds to the number of HXer units and utilities. From our answer in part (a), we note that:

- Above the pinch, there are two stream matches and one hot utility; therefore $U = 2 + 1 = 3$.
- Below the pinch, there are two stream matches and one cold utility; therefore $U = 2 + 1 = 3$.

This corresponds to the values given in the table for U.

Finding the value of N

Let us now figure out the values of N. We know that by definition, N represents the total number of streams and utilities.

- Above the pinch, there are four streams and one hot utility; therefore $N = 4 + 1 = 5$.
- Below the pinch, there are three streams and one cold utility; therefore $N = 3 + 1 = 4$.

Finding the value of L

Next, we find out the values of L, i.e., the number of loops. By inspecting our grid diagram in part (a), we note that we do not have any loops both above and below the pinch respectively. Loops are typically created when we have repeated matching between the same pairs of streams, and this is not the case for the HEN design in part (a). Therefore, $L = 0$ for both the sections above and below the pinch.

Finding the value of S

Finally, we need to find out the number of subsystems, S. This is oftentimes less obvious to figure out directly, compared to the rest of the unknowns in Euler's equation. It is easier to find S using Euler's equation and the previously found values of N, L, and U as shown below.

$$U = N + L - S$$

$$S = N + L - U$$

$$S_{\text{above pinch}} = 5 + 0 - 3 = 2$$

$$S_{\text{below pinch}} = 4 + 0 - 3 = 1$$

The complete table of values is shown below.

Variables	Above the pinch	Below the pinch
U	3	3
N	5	4
L	0	0
S	2	1

The typical value of S is one for straightforward HEN designs; however it is not ideal to assume so especially when the system becomes more complex. In this case, we notice that the number of subsystems above the pinch is equivalent to two. We can explain this using the grid diagram in part (a), whereby the two subsystems above the pinch correspond to two thermally isolated subsets, whereby each subset fulfills its own thermal load internally and independent of the other subset.

We observe from our grid display that stream 1 and stream 4's thermal loads satisfy each other completely, thereby forming one subset, while the other subset consists of streams 2 and 3 which also similarly satisfy each other's thermal loads completely and internally.

Problem 17

The process flow diagram (PFD) for a chemical plant is shown below. The system consists of three reactors A, B, and C, as well as associated heaters and coolers marked as circular symbols "H" and "C" respectively. The system satisfies all its heating and cooling duties using hot and cold utilities only. The intermediate stream temperatures and heater and cooler duties are also shown in the diagram.

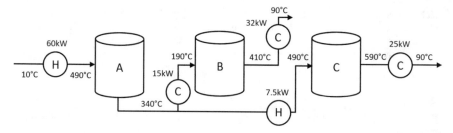

Given that the minimum approach temperature is 20 K, and assuming that heat capacities remain constant for all streams and that there is no phase change in the system,

(a) **Design a HEN design that achieves maximum energy recovery (MER) for this system.**

1. **Determine the pinch temperatures and minimum utility requirements and show your results in an energy cascade diagram.**
2. **Illustrate your HEN design for MER on a grid display for the sections above and below the pinch, and determine the value of U and the number of heat exchanger units required.**

(b) **Draw a process flow diagram for this system that incorporates the HEN design obtained in part (a).**

Solution 17

(a)(1)

Let us begin by identifying all discrete process streams in the overall system and labeling them as hot or cold streams. Note that a stream is hot if its target temperature is lower than its supply temperature (i.e., needs to be cooled), and vice versa for a cold stream.

We can calculate the flow rate heat capacities of the streams using the enthalpies (in kW) and temperatures indicated in the PFD. Note that there is a positive enthalpy change when the stream gains heat (e.g., over a heater) and a negative enthalpy change when a stream is cooled.

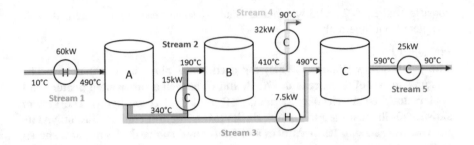

Description (color-coded in diagram above)	Stream label	Supply temperature T_{supply} [°C]	Target temperature T_{target} [°C]	Hot/ cold stream?	Enthalpy change ΔH [kW]	Heat capacity W [kW/K]
Reactor A feed	1	10	490	Cold	60	$\frac{60}{490-10} = 0.125$
Reactor A product to Reactor B feed	2	340	190	Hot	−15	$\frac{15}{340-190} = 0.1$
Reactor A product to Reactor C feed	3	340	490	Cold	7.5	$\frac{7.5}{490-340} = 0.05$
Reactor B product	4	410	90	Hot	−32	$\frac{32}{410-90} = 0.1$
Reactor C product	5	590	90	Hot	−25	$\frac{25}{590-90} = 0.05$

In order to design a HEN for MER, we should first find the pinch point and minimum heating and cooling utilities required by following the usual steps as summarized below:

1. Convert actual temperatures to shifted temperatures T^* using $\Delta T_{min} = 20$ K.
2. Tabulate shifted temperature intervals and calculate corresponding enthalpy change and cumulative enthalpy for each interval.

   (a) Identify the pinch point by inspecting the cumulative enthalpies.
   (b) Determine the minimum heating utility duty, $Q_{H,\,min}$.

3. Draw an energy cascade diagram to determine the minimum cooling utility duty, $Q_{C,\,min}$.

Let us now apply the above steps to our problem.

Step 1: Finding shifted temperatures
Given that $\Delta T_{min} = 20$ K, we can find shifted temperatures as follows:

- Add $\frac{1}{2}\Delta T_{min}$ to all cold stream temperatures.
- Subtract $\frac{1}{2}\Delta T_{min}$ from all hot stream temperatures.

Stream label	Supply temperature T_{supply} [°C]	Target temperature T_{target} [°C]	Hot/cold stream?	T_s^* [°C]	T_T^* [°C]	Heat capacity W [kW/K]
1	10	490	Cold	10 + 10 = 20	490 + 10 = 500	0.125
2	340	190	Hot	340 − 10 = 330	190 − 10 = 180	0.1
3	340	490	Cold	340 + 10 = 350	490 + 10 = 500	0.05
4	410	90	Hot	410 − 10 = 400	90 − 10 = 80	0.1
5	590	90	Hot	590 − 10 = 580	90 − 10 = 80	0.05

Step 2: Tabulating information for shifted temperature intervals

T^* interval [°C]	Stream (s) present	ΔT [°C]	Net heat capacity, $\sum W_C - W_H$ [kW/K]	$\Delta H = \Delta T$ $(\sum W_C - W_H)$	$\sum - \Delta H$ [kW]
580–500	5	580 − 500 = 80	−0.05	−0.05 (80) = −4	4
500–400	1, 3, 5	500 − 400 = 100	0.125 + 0.05 − 0.05 = 0.125	0.125 (100) = 12.5	4 − 12.5 = −8.5
400–350	1, 3, 4, 5	400 − 350 = 50	0.125 + 0.05 − 0.1 − 0.05 = 0.025	0.025 (50) = 1.25	4 − 12.5 − 1.25 = −9.75 (most negative)
350–330	1, 4, 5	350 − 330 = 20	0.125 − 0.1 − 0.05 = −0.025	−0.025 (20) = −0.5	4 − 12.5 − 1.25 + 0.5 = −9.25
330–180	1, 2, 4, 5	330 − 180 = 150	0.125 − 0.1 − 0.1 − 0.05 = −0.125	−0.125 (150) = −18.75	4 − 12.5 − 1.25 + 0.5 + 18.75 = 9.5
180–80	1, 4, 5	180 − 80 = 100	0.125 − 0.1 − 0.05 = −0.025	−0.025 (100) = −2.5	4 − 12.5 − 1.25 + 0.5 + 18.75 + 2.5 = 12
80–20	1	80 − 20 = 60	0.125	0.125(60) = 7.5	4 − 12.5 − 1.25 + 0.5 + 18.75 + 2.5−7.5 = 4.5

Most negative value for $\sum - \Delta H$

We can find the pinch point by finding the temperature interval where we have the **most negative value for $\sum - \Delta H$**. This occurs at the interval 400–350 when $\sum - \Delta H = -9.75 kW$. This means that

- The pinch temperature (shifted) is $T^*_{pinch} = 350\,°C$. We can convert this shifted temperature back to the actual hot and cold stream temperatures at the pinch, $T_{pinch,hot} = 360\,°C$ and $T_{pinch,cold} = 340\,°C$.
- The minimum hot utility required is $Q_{H,min} = 9.75$ kW.

Using the values in the table above, we can draw an energy cascade (shown below) for our system to determine the remaining unknown, i.e., the minimum external cooling utility, $Q_{C,min}$ required for MER.

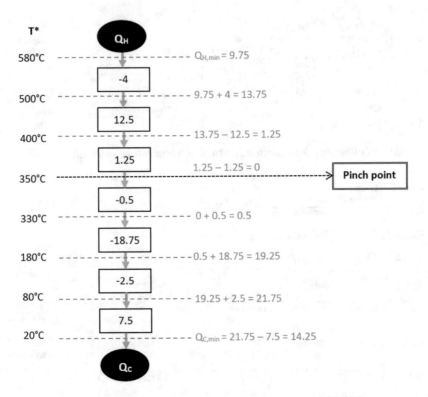

From the energy cascade, we find the value of $Q_{C,min} = \textbf{14.25 kW}$.

(2)

Let us design a heat exchanger network (HEN) for MER using a grid display. In a grid display, we indicate the supply and target temperatures of all streams (hot and cold), as well as the pinch temperatures. Stream matches can then be drawn using vertical lines that connect a hot stream to a cold stream. External utilities can also be placed on specific streams.

Let us proceed step-by-step, beginning with the section above the pinch.

Above the Pinch (Hot End)

Since the pinch point is at $T^*_{pinch} = 350\,°C$, the streams that are present above the pinch are those that have temperatures $T^* \geq 350\,°C$. From the earlier table, we identify the streams present in the section above the pinch are streams 1, 3, 4, and 5. Note that stream 2 is absent above the pinch, since it only ranges from T_s^* of 330 C to T_T^* of 180 C.

Let us draw the four streams above the pinch and indicate their respective heat capacities as shown below.

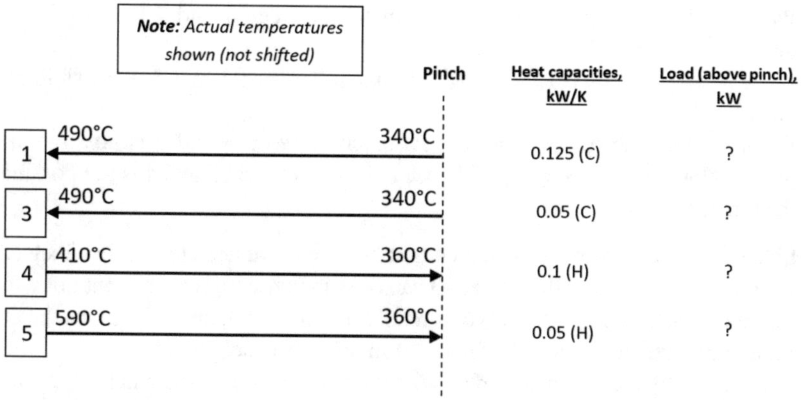

Before we can place stream matches, we need to calculate the respective heating and cooling loads of the streams above the pinch.

Stream	Heat capacity [kW/K]	Load (above pinch) [kW]
1	0.125	$0.125(490 - 340) = 18.75$
3	0.05	$0.05(490 - 340) = 7.5$
4	0.1	$0.1(410 - 360) = 5$
5	0.05	$0.05(590 - 360) = 11.5$

We can now update our grid display shown below with the thermal loads calculated in the table above.

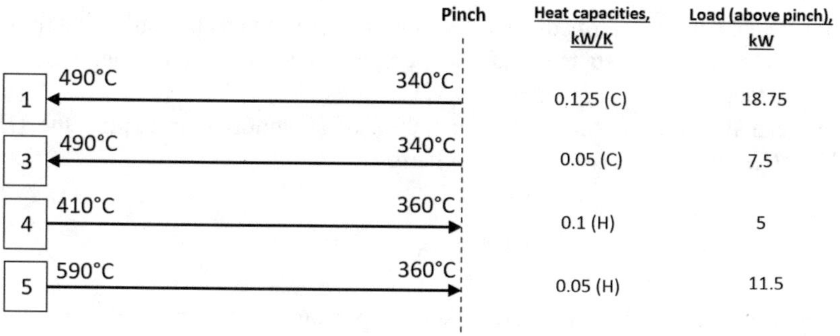

Let us now decide which stream matches to place. As a rule of thumb, we always **start placing matches from the pinch**, and move away from the pinch as we go along. For the section above the pinch, we need to check that $N_C \geq N_H$. Since we have two cold streams, $N_C = 2$, and two hot streams, $N_H = 2$, we fulfill the equality sign of the condition (i.e., $N_C = N_H$).

The next condition we need to check for matches at the pinch (hot side) is that $W_C \geq W_H$. In this case, we note the following matches fulfill this condition:

- $W_{C,\,1} \geq W_{H,\,4}$ since $0.125 \geq 0.1$: A match between 1 and 4 is possible at the pinch.
- $W_{C,\,1} \geq W_{H,\,5}$ since $0.125 \geq 0.05$: A match between 1 and 5 is possible at the pinch.
- $W_{C,\,3} \geq W_{H,\,5}$ since $0.05 = 0.05$: A match between 3 and 5 is possible at the pinch. [Note that $W_{C,\,3} < W_{H,\,4}$; hence a match between 3 and 4 is not possible at the pinch.]

Since there are three possible matches at the pinch, we need to decide which is the best. In this case, we note that stream 3 can only match with stream 5 but not with 4; therefore it makes sense to "reserve" stream 5 for stream 3, and this leaves the other two remaining streams 1 and 4 which can match with each other.

Let us now update our grid display with the two matches at the pinch, i.e., a match between streams 1 and 4 involving a heat exchanger (HXer) unit of 5 kW duty and a match between streams 3 and 5 involving a HXer unit of 7.5 kW duty.

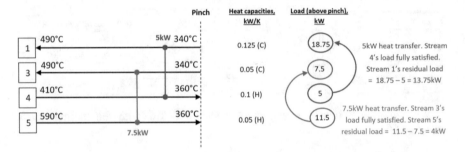

We next calculate intermediate temperatures of the streams that still carry residual loads. "Intermediate temperatures" are temperatures of streams after they have passed through the stream matches (or HXer units).

We can find the intermediate temperature of stream 1 upon exiting the HXer (5 kW HXer marked in red), as shown here.

$$340 + \frac{5}{0.125} = 380^\circ C$$

Similarly, we can find the intermediate temperature of stream 5 at the inlet of the HXer unit (7.5 kW HXer marked in dark yellow), as shown here.

$$360 + \frac{7.5}{0.05} = 510^\circ C$$

We will now update our grid display with the intermediate temperatures found above, and work on the next match based on residual loads that still need to be fulfilled. Note that all streams have now moved away from the pinch. Therefore, for the subsequent matches, it is not necessary to obey $W_C \geq W_H$.

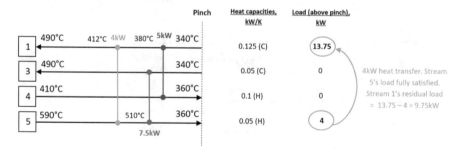

The next logical match is between streams 1 and 5 since both still have residual loads. This match will completely fulfill stream 5's load, and stream 1 will be left with a residual load of 9.75 kW. Like before, we are also able to determine the intermediate temperature of stream 1 after placing this match. This intermediate temperature is also the outlet temperature of stream 1 from the HXer representing this match.

$$380 + \frac{4}{0.125} = 412°C$$

This intermediate temperature is indicated together with the third match (4 kW HXer marked in blue) in the grid display below.

The residual load on stream 1 cannot be fulfilled by any further stream matches since all other streams have fulfilled their loads completely. Therefore, we will simply add an external hot utility of 9.75 kW to stream 1 which will completely fulfill its residual load. This completes our HEN design above the pinch.

All our results are consolidated in the grid display for the section above the pinch as shown below. We note that there are three HXers (corresponding to the three stream matches) and one external hot utility above the pinch.

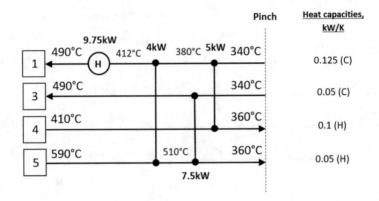

Below the Pinch (Cold End)

Like before, we first identify the streams that are present in the section below the pinch. Since our pinch occurs at $T^*_{pinch} = 350\,^\circ C$, streams that are present below the pinch must have shifted temperatures $T^* \leq 350\,^\circ C$. This means streams 1, 2, 4, and 5 are below the pinch. Note that stream 3 is absent below the pinch since it only ranges from T_s^* of $340\,^\circ C$ to T_T^* of $490\,^\circ C$.

Let us now draw these four streams below the pinch and indicate their respective heat capacities as shown below. It is worth noting that stream 2 does not reach pinch point.

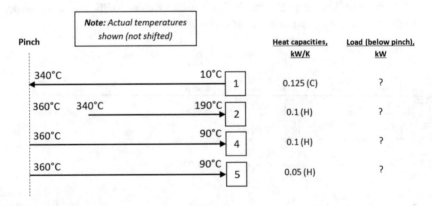

Before we place stream matches, let us calculate the respective heating and cooling loads of the streams below the pinch. This is shown below.

Stream	Heat capacity [kW/K]	Load (below pinch) [kW]
1	0.125	0.125(340 − 10) = 41.25
2	0.1	0.1(340 − 190) = 15
4	0.1	0.1(360 − 90) = 27
5	0.05	0.05(360 − 90) = 13.5

We may now update our grid display below with the thermal loads calculated in the table above.

Pinch	Note: Actual temperatures shown (not shifted)		Heat capacities, kW/K	Load (below pinch), kW
340°C		10°C [1]	0.125 (C)	41.25
360°C 340°C		190°C [2]	0.1 (H)	15
360°C		90°C [4]	0.1 (H)	27
360°C		90°C [5]	0.05 (H)	13.5

Since we are now working on the section below the pinch, we need to consider conditions that need to be fulfilled accordingly.

As before, we start placing matches from the pinch, and move away from the pinch as we go along. Below the pinch, we need to check that $N_H \geq N_C$ is always true. Since we have one cold stream, $N_C = 1$, and three hot streams, $N_H = 3$, we do fulfill the condition $N_H \geq N_C$.

The next condition we need to check for matches at the pinch (cold side) is $W_H \geq W_C$. In this case, we note that no matches obey the required condition due to the large value of $W_{C, 1} = 0.125$ kW/K.

- $W_{H, 2} \leq W_{C, 1}$ since $0.1 \leq 0.125$: A match between 1 and 2 is NOT possible at the pinch.
- $W_{H, 4} \leq W_{C, 1}$ since $0.1 \leq 0.125$: A match between 1 and 4 is NOT possible at the pinch.
- $W_{H, 5} \leq W_{C, 1}$ since $0.05 \leq 0.125$: A match between 1 and 5 is NOT possible at the pinch.

We can overcome this problem by splitting cold stream 1 into two separate streams, such that the original heat capacity value of 0.125 is then split into two smaller values as well. Let us label the two split streams as 1a and 1b and denote x as the heat capacity of one of the split streams (e.g., 1a) such that $W_{C, 1a} = x$. The heat capacity of the other split stream 1b would then have to be $W_{C, 1b} = 0.125 - x$.

Assuming stream 1a matches with stream 5 ($W_{H, 5} = 0.05$), then in order to fulfill the required condition $W_H \geq W_C$ at the pinch, we require $W_{C,1a} \leq 0.05$. The remaining unmatched cold stream 1b can then match with either 2 or 4 since both have the same heat capacity value ($W_{H, 2} = W_{H, 4} = 0.1$). And for this next match at the pinch, the condition $W_H \geq W_C$ requires that $W_{C,1b} \leq 0.1$.

Substituting $W_{C,\,1a} = x$ and $W_{C,\,1b} = 0.125 - x$ found earlier, we obtain

$$W_{C,1a} \le 0.05$$

$$x \le 0.05 \cdots \boxed{1}$$

$$W_{C,1b} \le 0.1$$

$$0.125 - x \le 0.1$$

$$x \ge 0.025 \cdots \boxed{2}$$

Combining both conditions $\boxed{1}$ and $\boxed{2}$ for the value of x, we have

$$0.025 \le x \le 0.05$$

Since there is a range of values that x can take, we may adopt the following tip when choosing a particular value:

- A good value to try for x is such that we use the lower limit of the range for $W_{C,\,1a}$, i.e., $x = 0.025$. This gives $W_{C,\,1b} = 0.125 - x = 0.1$. The sum of heat capacities for both split streams 1a and 1b should be equivalent to that of the original stream 1, i.e., $0.1 + 0.025 = 0.125$.

The updated tabulation of thermal loads for the split streams 1a and 1b is shown below.

Stream	Heat capacity [kW/K]	Load (below pinch) [kW]
1a	0.025	$0.025(340 - 10) = 8.25$
1b	0.1	$0.1(340 - 10) = 33$

The grid display is also updated with the split streams and their respective loads.

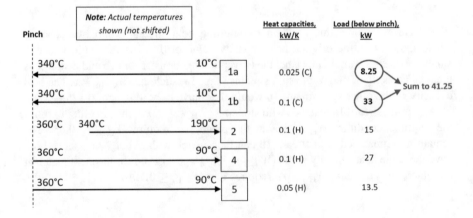

We have now resolved the earlier issue of being unable to fulfill the condition $W_H \geq W_C$ (at the pinch, cold side) by stream splitting. We are now ready to proceed with placing matches below the pinch, starting from the pinch point.

Stream 1a

- $W_{C, 1a} \leq W_{H, 2}$ since $0.025 \leq 0.1$: A match between 1a and 2 is possible at the pinch.
- $W_{C, 1a} \leq W_{H, 4}$ since $0.025 \leq 0.1$: A match between 1a and 4 is possible at the pinch.
- $W_{C, 1a} \leq W_{H, 5}$ since $0.025 \leq 0.05$: A match between 1a and 5 is possible at the pinch. [Note that $W_{C, 1b} > W_{H, 5}$; hence a match between 1b and 5 is not possible at the pinch.]

Since stream 5 can only match with stream 1a and not with 1b, it makes sense to "reserve" stream 1a for stream 5 at the pinch. This leaves the other remaining unmatched streams 1b, 2, and 4.

Stream 1b

- $W_{C, 1b} \leq W_{H, 2}$ since $0.1 \leq 0.1$: A match between 1b and 2 is possible at the pinch.
- $W_{C, 1b} \leq W_{H, 4}$ since $0.1 \leq 0.1$: A match between 1b and 4 is possible at the pinch.

As seen above, there are two matches at the pinch for stream 1b that fulfill $W_H \geq W_C$. In general, when we have more than one possible match, we can further select the best match as the one that **satisfies the full loads of streams as much as possible**. In this case, both matches are able to fully satisfy one of the stream's loads in the pair. [For the match between 1b and 2, stream 2's 15 kW load will be completely fulfilled, while the match between 1b and 4 will completely fulfill stream 4's 27 kW load.]

In this instance, it is then best to pick the match that fulfills the **largest** load so that we minimize any residual load. Hence, we proceed with the next match at the pinch between stream 1b and 4.

We now update our grid display with the two matches at the pinch (cold side) as mentioned above, i.e., between streams 1a and 5 and between streams 1b and 4. We can determine the HXer duties for the two matches as 27 and 8.25 kW as shown below.

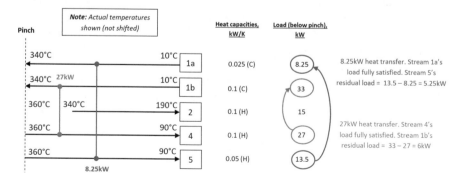

We may now proceed to calculate intermediate temperatures for any remaining streams with residual loads, which are streams 1b and 5. The intermediate temperature of stream 1b at the inlet of the HXer (marked in dark yellow) is

$$340 - \frac{27}{0.1} = 70\overset{\circ}{C}$$

The intermediate temperature of stream 5 at the outlet of the HXer (marked in red) is

$$360 - \frac{8.25}{0.05} = 195\overset{\circ}{C}$$

We can now update our grid display below with the intermediate temperatures found above and work on the next match.

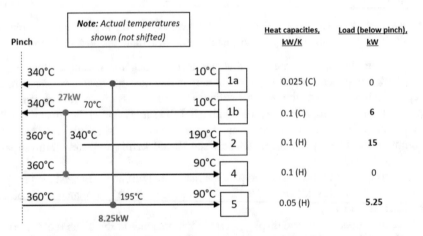

We observe that there are unfulfilled loads for streams 1b, 2 (full load), and 5. Noting that **we can only place cold utilities below the pinch** and not hot utilities, this means the **unfulfilled load for cold stream 1b must be fulfilled completely by matches only**. Cold streams can only match with hot streams (i.e., 1b and 2 or 1b and 5). In this case, we see that **cold stream 1b can only match with stream 2** since stream 5 will not be able to fulfill 1b's load completely. [Note that all streams have now moved away from the pinch; therefore it is not necessary to obey $W_C \geq W_H$ for this match and subsequent matches.]

With the third match between streams 1b and 2, the intermediate temperature of stream 2 at the outlet of the HXer (marked in blue) can be found as follows.

$$340 - \frac{6}{0.1} = 280\,^{\circ}C$$

The grid display below is updated with the third match as shown below.

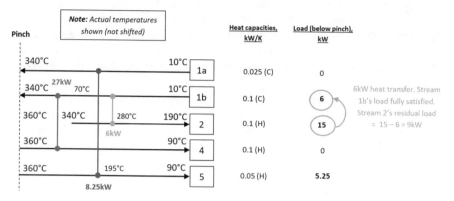

After the third match, there will only be unfulfilled loads remaining on hot streams 2 and 5. Their residual loads cannot be fulfilled by any further matches since there are no available cold streams left to match with (all cold streams have completely fulfilled their loads by previous matches). So, we complete our HEN design by placing external cold utilities of 9 and 5.25 kW on streams 2 and 5, respectively, to fulfill their residual loads completely.

The final grid display for the section below the pinch is shown below. We note that there are 3 HXer units (corresponding to the 3 stream matches) and 2 external cold utilities below the pinch.

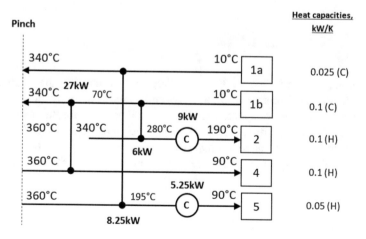

Combining our results for the sections above and below the pinch, we note that in total, we require **six HXers** for stream matches (inter-stream heat transfer) and **three external utilities** (two cold utilities and one hot utility). This gives us a value of

$U = 6 + 3 = 9$, for the total number of HXer units required for our HEN design for MER.

(b)

In a process flow diagram that incorporates a HEN design, we need to show how the streams flow through the six HXers and three utilities (for the HEN design in part (a)) and their connection with the process units of the overall system (i.e., Reactors A, B, C and their associated heaters and coolers). We should also indicate the intermediate temperatures of the streams as they enter and exit the HXers, utilities, and process units. This is shown in the PFD below, which is color-coded to better differentiate the streams.

In the diagram below, the six HXers for stream matches (inter-stream heat transfer) are shown as clear circles with the HXer duties indicated within. The three external utilities added to streams, i.e., the hot utility of 9.75 kW and two cold utilities of 9 and 5.25 kW, are shown as shaded circles (red for heating and blue for cooling).

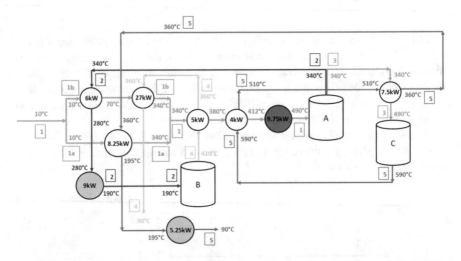

One point to note is that the heaters and coolers indicated in the original PFD actually comprise of a combination of HXer units/utilities whereby the sum total of their duties would achieve the required heating/cooling loads for the streams.

For example, for stream 1, the 60 kW heater shown in the original PFD actually provides heat via a combination of five HXers and one external hot utility as shown below marked in red. Their total duties then sum to the total 60 kW duty of the heater.

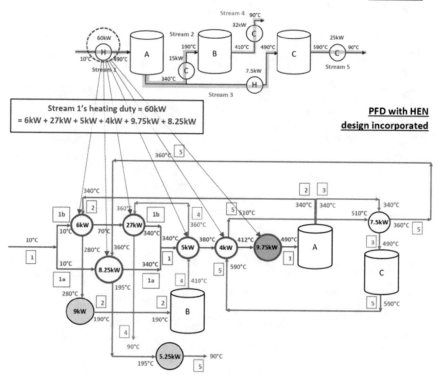

Stream 1's heating duty = 60kW
= 6kW + 27kW + 5kW + 4kW + 9.75kW + 8.25kW

Index

C

Capital cost (HEN optimization), 3, 4

Cold composite curves, 32, 43, 45–48, 80, 103, 104

Cold (cooling) utility, 3, 5, 9, 10, 13–15, 27, 30, 31, 35, 38, 41, 42, 44, 48, 51, 53, 58, 64–67, 69–71, 76–80, 82, 85, 89, 91–94, 96–99, 102, 105–107, 109, 120–122, 124, 126, 127, 129–133, 142–144

Constant temperature utility, 89, 91, 99

Coolant utility/fluid, 69, 89, 91–93, 97, 106, 107

Cost-benefit analysis, 2

Cross-pinch heat leakage/transfer, 31

Cumulative enthalpy, 34, 44, 53, 64, 80, 104, 132

Cyclic matching, 110, 113, 114, 117, 119, 120

D

Design equation (heat exchanger), 7, 8, 12, 48

E

Energy cascade, 19–62, 64, 64, 65, 80, 80, 102, 104, 105, 131, 132, 133, 134

Euler's network theorem/equation, 63, 69, 71, 109, 110, 119–121, 124–127, 129

F

Flow rate heat capacity, 1–3, 5, 8, 21, 22, 32, 43, 44, 46, 49, 52–54, 59, 65, 66, 100, 102, 110, 117, 121, 123, 128, 131

G

Grand composite curve (GCC), 64, 80, 81, 83, 84, 89, 91–94, 96–102, 104–108, 116

Graph theory/diagram, 121, 124–126

H

Heat integration, 1–17, 20, 22, 31, 35, 52, 60, 62, 72, 76, 85, 96

Heat sink, 1, 2, 20, 27, 31, 54, 80, 89, 102, 127

Heat source, 1, 20, 27, 31, 54, 87, 89, 91, 96, 102, 127

Heat transfer coefficient (heat exchanger), 7, 48, 87

Hot composite curve, 32, 45–48, 103, 104

Hot (heating) utility, 5, 9, 14, 15, 27–30, 32, 34, 35, 41, 42, 44, 48, 51, 53, 54, 59–62, 64, 67–71, 76–83, 85–89, 91, 94, 102, 105, 121, 122, 124, 126, 127, 129, 130, 132, 133, 137, 142–144

I

Inter-process heat transfer, 9

L

Latent heat, 83, 87, 89, 91, 97–99
Log mean temperature difference/average, 8
Loops, 63, 70–73, 110, 119–127, 130

M

Maximum energy recovery (MER), 7, 11, 16,
         32, 35, 42, 43, 51, 52, 59, 63–65, 69–73,
         82, 85, 101, 102, 109, 110, 117, 119,
         127, 131–134, 144
Minimum approach temperature, 7, 12, 17, 20,
         30, 32, 43, 48, 52, 59, 74, 102, 103, 109,
         127, 128, 131
Minimum utility duty, 11, 20, 38, 48–52, 102

N

Nose (Grand composite curve), 83–85, 96

P

Pinch analysis, 19–62, 64, 71
Pinch point, 7, 12, 17, 20, 27, 29, 31, 32, 34, 44,
         48–51, 53, 64, 72, 81, 102–104, 106,
         132–134, 138, 141
Pinch temperature (hot, cold), 15, 29, 34, 38,
         40, 41, 52, 59, 60, 67, 75, 128
Pocket (Grand composite curve), 85

R

Raising steam, 89, 93, 96, 97

S

Sensible heat, 83, 87, 91, 92, 97, 98, 106
Shifted temperature, 32–34, 43, 44, 52, 53, 64,
         81, 82, 86, 102–104, 116, 132, 133, 138
Steam levels, 89, 94–96, 99, 100
Stream-splitting, 55, 57–59, 109, 110, 117, 119
Subsystem/subnetwork, 32, 48, 63, 70, 96, 102,
         109, 110, 120, 123, 125–127, 130
Supply temperature, 5, 7, 21, 33, 38, 40, 42, 43,
         59, 68, 77, 78, 115, 123, 131

T

Target temperature, 1–5, 7–10, 15, 21, 22, 33,
         36, 37, 41–43, 45, 48, 49, 51, 53, 59,
         62, 64, 67, 72, 79, 80, 111, 117, 123,
         131, 134
Temperature against enthalpy, 8, 45, 80
Thermodynamics (laws), 11, 13, 15, 16, 73, 74,
         79–81, 99, 108

U

Utility cost (HEN optimization), 3

V

Variable temperature utility, 84, 89, 91–93,
         97, 98

Printed in the United States
by Baker & Taylor Publisher Services